Collins

Revision

NEW GCSE

Biology

OCR

Twenty First Century Science

Authors: **Eliot Attridge**
John Beeby

Revision Guide +
Exam Practice Workbook

Contents

Contents

What genes do

What genes are and how they work

- **Genes** carry the instructions that control how you develop and function. They do this by telling the cells to make the **proteins** needed for your body to work.

- Each gene is a section of a very long molecule of a chemical called **DNA** (deoxyribonucleic acid).

- Lengths of DNA are coiled and packed into structures called **chromosomes**. These are found in the nuclei of the body's cells. We have between 20 000 and 25 000 genes on our chromosomes.

- Strands of DNA are made up of four chemicals called bases, as well as phosphate groups and sugar molecules.

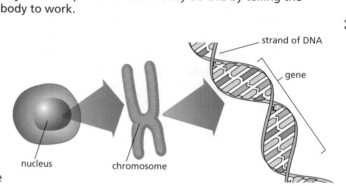

A gene is a length of DNA that codes for a particular protein.

- The order of the bases in a DNA strand determines the order of amino acids in a protein.

- Proteins fall into two groups:

  1 **Functional proteins** enable the body to function. Examples include **enzymes**, antibodies and hormones.

  2 **Structural proteins** give the body structure, rigidity and strength. Examples include collagen in ligaments and keratin in skin.

- The Human Genome Project has identified the location of all the genes on human chromosomes. We call the complete gene set of an organism its genome.

- The project will help us to understand how genes control our characteristics and development, and can lead to certain diseases.

- The project has ethical implications, e.g. some drug companies want to patent or 'own' genes. They could then charge other scientists money to investigate the genes, which would restrict research.

We're all different

- Our characteristics are controlled by our genes (e.g. dimples), our **environment** (e.g. the presence of scars or dyed hair), or a combination of these (e.g. our body weight).

- We inherit genes from each of our parents, so we're similar, but not identical to, each parent.

- Differences in genes produce **variation** in offspring.

- Some characteristics are controlled by several genes working together. These characteristics will show **continuous variation** across a population, e.g. the continuous range of eye colours and different heights.

- For a particular characteristic, you can describe a person by their **genotype** or **phenotype**.

- A person's genotype is his or her genetic make-up. It can refer to the whole of an individual's genes or (more often) the genes for a particular characteristic, such as if we have dimples. The genotype is usually written as two letters, for example DD.

- A person's phenotype describes a his or her observable physical features, e.g. build, or a single characteristic. The phenotype will depend on the person's genes, but may also be affected by how these interact with the environment.

Studies on twins

- Identical twins have identical genotypes because they develop after a fertilised egg splits into two.

- Studies of identical twins, especially those that have been separated, can help us to understand the effect of the environment on a person's genotype.

> **Remember!**
>
> Remember, the order of size, from largest to smallest, is:
>
> cell → nucleus → chromosomes → DNA → gene → base

Improve your grade

Studies on twins

Higher: How do studies of identical twins help us to understand the effect of the environment on the phenotype for a characteristic?

AO2 [3 marks]

Genes working together and variation

How genes and chromosomes are organised

- Chromosomes can be arranged into pairs. Human cells have 23 pairs of chromosomes (a total of 46 chromosomes).
- Sex cells are eggs (ova) and sperm. These have 23 chromosomes, one from each pair.
- At **fertilisation**, an egg and sperm join together to produce a **zygote**. The zygote has a full set of 46 chromosomes – 23 from the mother and 23 from the father.
- The human baby that develops has a combination of genes from its mother and father.
- Pairs of chromosomes have genes for the same characteristic at identical positions on each chromosome of the pair.

- Changes to our DNA sometimes occur, causing a **mutation**. This can take place when sex cells are being made, or after fertilisation.
- One type of mutation is a chromosome mutation. This results in an individual having extra chromosomes. For example, a person with an extra chromosome 21 will have Down's syndrome.

Variation in offspring

- The combination of chromosomes in an egg or sperm will always be different. For example, in an egg, chromosome 1 could have been inherited from the mother, while chromosomes 2 and 3 could have been inherited from the father, etc. So, the combination of chromosomes (and therefore genes) will be unique to that person – unless he or she is an identical twin.
- Environmental effects will also add to the variation.

Pairs of alleles

- **Alleles** are the different forms in which the genes controlling a characteristic can occur. So, for dimples: one allele is for the presence of dimples and one allele is for a lack of dimples.

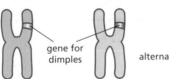

gene for dimples

alternative alleles: ☐ dimples ☐ no dimples

The alleles for having dimples or for not having dimples are located on the same part of each chromosome in a pair. This individual has one of each allele.

- If the two alleles of a gene are identical, the person is said to be **homozygous** (for that characteristic).
- If the two alleles are different, the person is said to be **heterozygous** (for that characteristic).

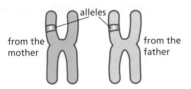

alleles

from the mother

from the father

This pair of chromosomes shows the same pair of alleles, so this individual is homozygous. In the diagram above the individual is heterozygous.

Remember!

Two combinations of alleles, e.g. DD and dd, are possible in an organism that is homozygous for a characteristic. Only one combination, i.e. Dd, is possible for an organism that is heterozygous for a characteristic.

Higher

Ideas about science

You should be able to:

- distinguish between questions that could be answered using a scientific approach (e.g. determining which alleles are dominant and which are recessive) and questions which are difficult to answer or could not be answered (e.g. the combination of alleles that occurs when sex cells are produced and at fertilisation).

Improve your grade

Variation in offspring

Foundation: Explain how variation occurs in the offspring produced by humans.

AO1 [5 marks]

Genetic crosses and sex determination

Genetic traits

- Traits are passed on from parents to their offspring through genes on chromosomes.
- Genes for a particular trait are found at the same place on each chromosome of the chromosome pair.
- The different forms of a gene that control a certain trait are called alleles.
- In most cases, alleles for a trait can be **dominant** or **recessive**. For example, the allele for hairy toes is dominant to the allele for hairless toes.
- Dominant alleles are written with upper-case letters in genetic diagrams, e.g. H for hairy toes.
- Recessive alleles are written with lower-case letters in genetic diagrams, e.g. h for hairless toes.
- If at least one dominant allele is present (e.g. Hh), the trait shown will be the dominant one (e.g. hairy toes).
- For the recessive trait (e.g. hairless toes) to be shown there must be two recessive alleles present (e.g. hh).

Genetic diagrams

- You can use a **Punnett square** to:
  - show genetics crosses
  - find the probability of two parents producing different types of offspring.
- You can show the inheritance of a trait in a family over several generations using a **family tree diagram**.
- A family tree diagram is very useful when tracing a genetic disorder, such as Huntingdon's disease, over generations.

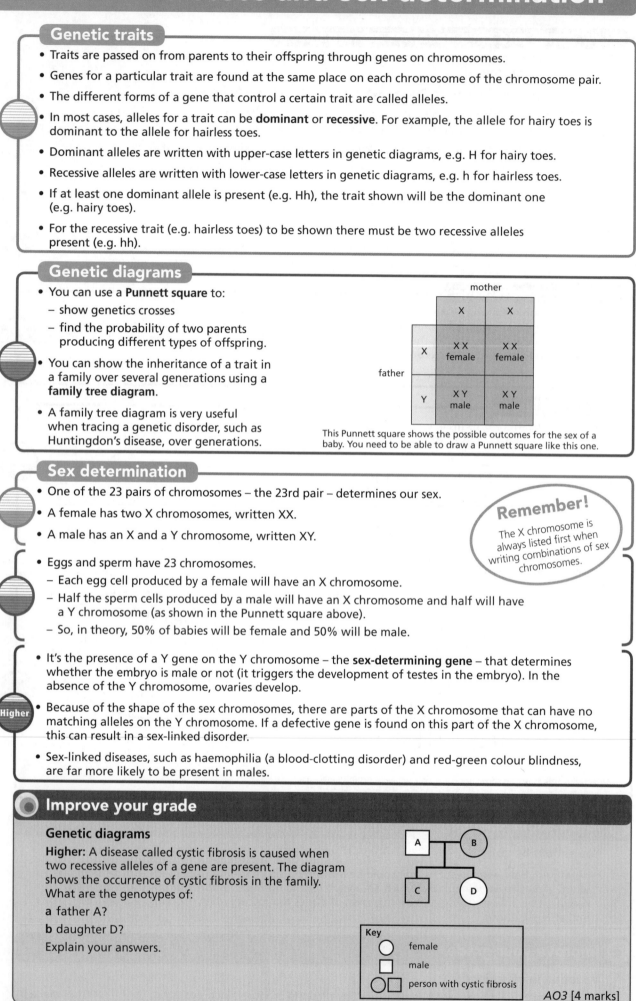

This Punnett square shows the possible outcomes for the sex of a baby. You need to be able to draw a Punnett square like this one.

Sex determination

- One of the 23 pairs of chromosomes – the 23rd pair – determines our sex.
- A female has two X chromosomes, written XX.
- A male has an X and a Y chromosome, written XY.

- Eggs and sperm have 23 chromosomes.
  - Each egg cell produced by a female will have an X chromosome.
  - Half the sperm cells produced by a male will have an X chromosome and half will have a Y chromosome (as shown in the Punnett square above).
  - So, in theory, 50% of babies will be female and 50% will be male.

- It's the presence of a Y gene on the Y chromosome – the **sex-determining gene** – that determines whether the embryo is male or not (it triggers the development of testes in the embryo). In the absence of the Y chromosome, ovaries develop.

- **Higher:** Because of the shape of the sex chromosomes, there are parts of the X chromosome that can have no matching alleles on the Y chromosome. If a defective gene is found on this part of the X chromosome, this can result in a sex-linked disorder.

- Sex-linked diseases, such as haemophilia (a blood-clotting disorder) and red-green colour blindness, are far more likely to be present in males.

> **Remember!**
> The X chromosome is always listed first when writing combinations of sex chromosomes.

Improve your grade

Genetic diagrams

Higher: A disease called cystic fibrosis is caused when two recessive alleles of a gene are present. The diagram shows the occurrence of cystic fibrosis in the family. What are the genotypes of:

a father A?

b daughter D?

Explain your answers.

Key

◯ female

▢ male

◯▢ person with cystic fibrosis

AO3 [4 marks]

Gene disorders, carriers and genetic testing

Gene disorders

- Some disorders are caused by defective or faulty alleles.

- Huntington's disease is a dominant disorder, i.e. the presence of a *single* dominant allele will cause the disease. It occurs in middle age. Symptoms include tremors (uncontrollable shaking), memory loss, inability to concentrate and mood changes.

- *Both* alleles on the chromosome pair must be recessive for a person to get a recessive disorder, such as cystic fibrosis. Symptoms include the production of thick gluey mucus that affects the lungs and makes digesting food difficult, breathing problems and chest infections.

- For recessive single gene disorders, a person with a normal and a defective allele will be normal (because the normal gene is dominant), but they will be a **carrier**.

- If both parents are carriers they can give birth to a child with the disease.

Remember!

Only one defective allele in a pair is needed to cause the disease in a dominant disorder. Both alleles need to be defective for a recessive disorder to occur.

Genetic testing

- **Genetic screening** is used to check for a particular disorder, even when there is no history of it in the family. It is hoped this will minimise the damage such disorders can cause. For example, the heel prick 'blood spot test' is used on most newborn babies to diagnose rare genetic disorders.

- **Genetic testing** of individuals is carried out when a genetic disease, such as cystic fibrosis, runs in the family. This may allow people to get treatment for the disease or to plan for the future.

- Genetic testing sometimes raises ethical questions. For example, if a person has Huntington's disease should they tell their employer, insurance company or family?

- Genetic testing during pregnancy may involve **cell sampling** by:
  - amniocentesis (collecting cells from the developing **fetus** which are present in amniotic fluid)
  - chorionic villus sampling (testing a sample of cells taken from the placenta).

  Both tests carry risk – around 1% of babies are miscarried as a result of genetic testing.

- Production of **embryos** by *in vitro* fertilisation (IVF) allows doctors to check the genetic make-up of the embryos prior to implantation. This is known as 'embryo screening'.

- Embryo screening is used to investigate families with a known history of a disorder, such as cystic fibrosis. It allows doctors to remove any embryos suffering from a disorder and implant only normal embryos.

- Screening embryos prior to implantation and only using healthy embryos is called **pre-implantation genetic diagnosis (PGD)**.

- Procedures like PGD and embryo research are carefully monitored in the UK. Guidelines for clinics and research centres cover ethical and moral considerations for embryo use.

- Parents likely to pass on a genetic abnormality:
  - may decide not to have a family
  - may need to decide on whether to continue with the pregnancy or terminate it.

- Types of genetic tests may give false negative results and may *sometimes* give false positive results (where the test is positive but the person does not have the disorder).

Higher

Ideas about science

You should be able to:

- show awareness of, and discuss, the official regulation of scientific research and the application of scientific knowledge, e.g. the regulation of embryo testing and research

- say clearly what the ethical issue is, e.g. the use of information from genetic testing, and summarise different views

- identify and develop arguments that consider the best outcome for the greatest number of people, e.g. the mass screening of newborn babies.

Improve your grade

Gene disorders

Foundation: Write down one dominant genetic disorder and one recessive genetic disorder. For each disorder, list three symptoms the person will show.

AO1 [4 marks]

Cloning and stem cells

Cloning

- **Clones** are individuals with identical genes.
- In **asexual reproduction** only one parent is involved, so the offspring has identical DNA to the parent. **Bacteria**, along with some plants and simple animals (e.g. *Hydra*), reproduce asexually.
- Plants can reproduce asexually by:
  - using runners (for example strawberries) – shoots sent out that grow into identical plants
  - producing bulbs (for example daffodils).
- Identical twins are human clones produced when a fertilised egg splits, resulting in two genetically identical individuals.

- As clones have identical DNA, any differences between individuals in a clone and their parent must be a result of the environment and *not* genes.
- The advantages of producing clones / asexual reproduction are that:
  - successful characteristics are seen in offspring
  - asexual reproduction is useful where plants and animals live in isolation.
- The disadvantage of producing clones / asexual reproduction is that there is no genetic variation. This means that if conditions change or there is disease, the population could be wiped out.

- Clones have been produced by artificial animal cloning, for example Dolly the sheep and Snuppy the dog.
  - The nucleus from a body cell is extracted and inserted into an egg cell that has had its nucleus removed. This gives the egg cell a full set of genes without having been fertilised.
  - The embryo is implanted into a suitable surrogate mother.
- It is illegal to create clones of humans in many countries, including the UK.

Stem cells

- A human embryo develops from a single cell. This cell divides over and over again as the baby develops. Most of these cells become specialised to do different jobs (a process called **differentiation**).
  - After five days, the embryo is a ball of cells containing **embryonic stem cells**.
  - These cells are unspecialised. They divide and develop into the different types of cell in the human body.
- As adults, some **stem cells** – **adult stem cells** – remain in certain parts of our bodies.
- Adult stem cells can repair or replace certain cell types. For example, bone marrow cells are able to develop into different types of blood cells.

- Adult stem cells are used to treat various diseases, but have limited uses.
- Because embryonic stem cells can develop into other cell types they have huge potential. However, their use is controversial:
  - They are usually taken from unused embryos following fertility treatments.
  - Their use involves the destruction of the embryo.
- Recent research is focusing on reprogramming adult body cells into stem cells, and collecting cells from the umbilical cord blood when a baby is born.

- Stem cells could be used in:
  - the testing of new drugs
  - understanding how cells become specialised in the early stages of human development by the switching on and off of particular genes
  - renewing damaged or destroyed cells in spinal injuries, heart disease, Alzheimer's disease and Parkinson's disease.

EXAM TIP

Make sure you understand:
- the difference between embryonic and adult stem cells
- the difference in potential that these have in the treatment of disease.

Improve your grade

Stem cells

Higher: Discuss the future use of stem cells in medicine.

AO1, AO2 [4 marks]

B1 Summary

Genes carry instructions to control how an organism develops.

The instructions tell the cell how to make essential proteins.

Proteins are functional, e.g. enzymes, or structural, e.g. collagen.

How genes control our characteristics and development

Human characteristics are determined by:

- genes, e.g. dimples
- the environment, e.g. scars
- or by a combination of both, e.g. body weight.

Many characteristics are controlled by several genes working together, e.g. eye colour.

Genes control how an organism functions.

They are found in the nucleus of the cell. They are sections of the DNA molecules that make up chromosomes.

Sex determination:

- In human females the sex chromosomes are XX.
- In human males the sex chromosomes are XY.

The Y chromosome has the sex-determining gene, which causes testes to develop. Without this (i.e. in females), ovaries develop.

Sexual reproduction causes variation as the offspring has genes from both parents. Offspring are similar to parents because of genes inherited from them. Siblings differ because they inherit different combinations of genes.

Genetic diagrams – Punnett squares and family trees – are used to show genetic crosses.

Genetics and inheritance

The genotype of an organism is its genetic make-up. The phenotype describes the organism's features.

Different forms of a gene are alleles. An individual:

- is homozygous when two alleles are the same for a characteristic
- is heterozygous when two alleles are different for a characteristic.

Alleles can be dominant or recessive. With one or two dominant alleles, the individual shows the characteristic. The recessive characteristic is seen only in individuals with two recessive alleles.

Body cells have pairs of chromosomes; sex cells have one chromosome from each pair.

Chromosomes have the same type of genes in the same place on each chromosome of the pair.

Single gene disorders are caused by faulty alleles of a gene.

The faulty gene can be dominant or recessive.

For recessive single gene disorders, a person with a single recessive gene will not have the disorder but will be a carrier.

A Punnett square or family tree shows the risk of inheriting a disorder/being a carrier.

Implications of testing for genetic diseases:

- tests carry risk of miscarriage
- results not 100% reliable
- decision as to whether to have children or abort a pregnancy
- should an employer, the family or insurance company be told?

Genetic diseases

Genetic testing is carried out when genetic disease runs in the family.

Genetic screening is carried out on a large scale, e.g. in newborns, where there is no history of a disease.

Embryos produced by IVF can be screened before implantation (pre-implantation genetic diagnosis – PGD).

PGD and embryo research are carefully monitored in the UK.

Natural clones are individuals with identical genes, so any differences between individuals must be due to environmental factors.

Some organisms reproduce asexually to produce clones.

Identical twins are formed when cells of an embryo separate.

Clones

Artificial clones can be produced when a nucleus from a body cell is transferred to an unfertilised egg.

Embryonic stem cells can develop into any type of specialised cell. Adult stem cells can develop into fewer cell types.

Being unspecialised, stem cells have potential in the treatment of disease.

Microbes and disease

How microorganisms cause disease

- **Microorganisms** that cause disease and make us feel ill are called **pathogens**. Pathogens include bacteria and viruses.

- When microorganisms get into the body, they reproduce quickly and cause symptoms of disease.

	Diseases caused by bacteria	Diseases caused by viruses
Symptoms caused by:	release of poisons or **toxins** by the bacteria	damage to the cells as the viruses reproduce
Examples:	bacterial meningitis, tetanus, salmonella food poisoning, tuberculosis (TB)	influenza (flu), the common cold, measles and chickenpox

- Bacteria reproduce by dividing into two, which is a type of asexual reproduction called **binary fission**.

- Bacteria reproduce rapidly (this is called exponential growth) in the ideal conditions of the human body.

- Viruses need a 'host' cell to reproduce. They enter the host cell and 'hijack' the cell's mechanisms for making DNA and proteins, and make copies of themselves.

Remember!
Viruses always need a cell to live in. They can't live outside the body for very long.

- The copies of the virus are released in very large numbers from the infected cell and go on to infect other cells and/or other people.

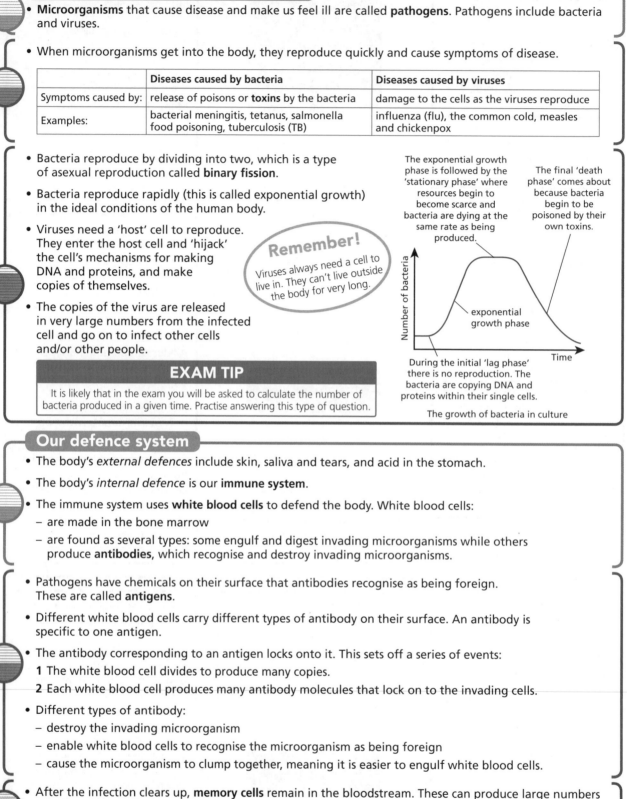

The exponential growth phase is followed by the 'stationary phase' where resources begin to become scarce and bacteria are dying at the same rate as being produced.

The final 'death phase' comes about because bacteria begin to be poisoned by their own toxins.

Number of bacteria

exponential growth phase

Time

During the initial 'lag phase' there is no reproduction. The bacteria are copying DNA and proteins within their single cells.

The growth of bacteria in culture

EXAM TIP

It is likely that in the exam you will be asked to calculate the number of bacteria produced in a given time. Practise answering this type of question.

Our defence system

- The body's *external defences* include skin, saliva and tears, and acid in the stomach.

- The body's *internal defence* is our **immune system**.

- The immune system uses **white blood cells** to defend the body. White blood cells:
  - are made in the bone marrow
  - are found as several types: some engulf and digest invading microorganisms while others produce **antibodies**, which recognise and destroy invading microorganisms.

- Pathogens have chemicals on their surface that antibodies recognise as being foreign. These are called **antigens**.

- Different white blood cells carry different types of antibody on their surface. An antibody is specific to one antigen.

- The antibody corresponding to an antigen locks onto it. This sets off a series of events:
  1 The white blood cell divides to produce many copies.
  2 Each white blood cell produces many antibody molecules that lock on to the invading cells.

- Different types of antibody:
  - destroy the invading microorganism
  - enable white blood cells to recognise the microorganism as being foreign
  - cause the microorganism to clump together, meaning it is easier to engulf white blood cells.

- After the infection clears up, **memory cells** remain in the bloodstream. These can produce large numbers of antibodies very quickly if the microorganism enters the body again.

- At this point the person is said to be **immune** to that particular pathogen.

Improve your grade

Our defence system

Higher: A bacterium enters your blood stream. Describe the series of events leading to the bacterium being destroyed by your immune system. *AO1* [5 marks]

B2 Keeping healthy

Vaccination

Vaccination programmes

- White blood cells make antibodies against chemicals on the surface of pathogens. These chemicals are called antigens.
- Memory cells left in the body can produce antibodies very quickly if they meet the living disease-causing microorganism.
- A **vaccine** contains a safe form of the microorganism that causes a disease, so that you don't become ill after receiving it.
- In the bloodstream, the immune system attacks the microorganism in the vaccine.

- Vaccination programmes protect children against diseases that are preventable.
- Babies and children undergo a course of **vaccinations** in their first year or so.
- Some pathogens do not change over time, so the same vaccines can be used against these, year after year.
- Other pathogens, such as the influenza (flu) virus, change rapidly, so new vaccines must be developed.

- An epidemic occurs if a disease spreads rapidly through a population, for example in a city or country.
- To avoid an epidemic, it is necessary to vaccinate a high percentage of the population – this leads to what is called 'herd immunity'.
- Widespread vaccination has eradicated one disease – smallpox – from the world. It has also reduced childhood diseases such as measles, mumps and rubella.
- The longer-term aim is to eradicate certain diseases altogether.

Higher

Making vaccination safe

- Scientists test new vaccines very carefully as they are developed to check for any **side effects**.
- Side effects can be more severe in some people than in others, because of genetic variation.
- No type of medical treatment can ever be *completely* risk-free (but people often think the risk is higher than it is).

- Vaccinations are *extremely* safe, and millions of people have benefited from them.
- *Occasionally*, a child may develop a minor adverse reaction (such as a rash or fever). In *very rare cases*, the reaction can be more serious. Any adverse reactions are recorded and followed up.
- Any *risks* of vaccination must be considered against the *benefits*. Any vaccine generating an unusual number of adverse reactions will be quickly withdrawn.

Other ways of reducing infection

- **Antimicrobials** are a group of substances that are used to kill microorganisms or inhibit (slow) their growth. They are effective against bacteria, viruses and fungi.
- **Antibiotics** are a type of antimicrobial which are effective against bacteria but not against viruses. They allow doctors to treat illnesses caused by bacteria, such as tuberculosis.
- Studies show that some microorganisms are developing **resistance** to these antimicrobials because of their very wide use.
  - Resistance to antimicrobials means that some strains of bacterial infections are now difficult to control.
  - Antimicrobial-resistant microbes can be a particular problem where antimicrobials are used frequently, such as in hospitals.

Improve your grade

Vaccination programmes

Higher: The graph shows the number of cases of measles, and deaths from measles, in England and Wales from 1940 to 2008.

Discuss what the data suggest about the effectiveness of the measles vaccines.

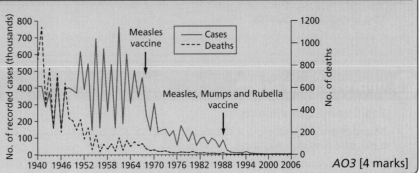

AO3 [4 marks]

Safe protection from disease

Antimicrobials

- Antimicrobials are chemicals that prevent the growth of microorganisms, so provide protection against disease. They include:
  - antibiotics against bacteria; antivirals against viruses; antifungals against fungi
  - many cleaning products, along with antiseptics and disinfectants.

- Over a period of time, bacteria and fungi can develop resistance to antimicrobials. In a population of microorganisms, some may be resistant to the antimicrobial. These will survive the use of the antimicrobial and pass on their resistance. The resistance spreads through the population of microorganisms.

- Resistant microorganisms are sometimes called '**superbugs**'.

- The overuse of certain antibiotics has led to some microorganisms becoming resistant to them. For this reason a course of antibiotics should be:
  - prescribed *only* for more serious infections, when they are really needed
  - completed, so that the bacteria causing the infection are killed completely.

resistance level

How the use of antimicrobials can lead to a more resistant population of microbes:
a initial population of microorganisms
b directly after treatment with antimicrobial
c final population of microorganisms

- Random changes in the genes – called **mutations** – give some bacteria resistance to antimicrobials.

- Because of the rapid reproductive rate of bacteria, antimicrobial-resistant genes spread through the population.

- Antibiotic resistance has led to some strains of bacteria that are very difficult to eradicate. These include MRSA, which is a problem in many hospitals.

Trialling new treatments

- New medicines, vaccines and other treatments are tested very carefully before being made available to the general public.

- Early stages of testing involve human cells grown in the laboratory, and animals.

- If the drug seems to be effective and safe, it is tested on humans in **clinical trials**. These are carried out on healthy volunteers (to check for safety) and on people with the illness (to test for safety and effectiveness).

- When drug trials are carried out, one group of people is given the new drug being trialled. The results are compared with those of a **control group**.

- One type of control group receives the existing treatment; another type of control group receives a **placebo** – a tablet or liquid made to look like the drug, but without the active ingredient.

- One ethical issue related to a drug trial is that it must not disadvantage the patient. If evidence from the trial suggests the new drug is effective, it is offered straightaway to patients receiving the placebo.

- In an 'open-label' trial, both researchers and patients know which drug the patient is receiving.

- In a 'blind' study, the patient doesn't know which drug they are receiving but the researcher does.

- In a 'double-blind' study, neither patient nor researcher knows which drug is being given.

- Some trials investigate the effects of the drug over a long period of time. This is because:
  - side effects may appear, or increase, over time
  - the drug may become less effective.

Ideas about science

You should be able to:

- evaluate critically the design of a study to test if the use of a new drug is effective in the treatment of disease.

Improve your grade

Trialling new treatments

Foundation: When testing a chemical on humans to see if it would be suitable as a new antibiotic, a placebo is sometimes used.

Explain why placebos are important, and how ethical issues with using placebos are overcome. *AO1* [4 marks]

The heart

The heart and circulatory system

- The heart, blood vessels and blood make up the **circulatory system**.
- The blood carries nutrients and oxygen to the body's cells, and removes and carries waste products from the cells.
- The blood is pumped around the body in blood vessels by the heart.
- The heart is a 'double pump' – as one half is pumping oxygenated blood from the lungs to the body, the other half is pumping deoxygenated blood from the body to the lungs.
- At the lungs, deoxygenated blood absorbs oxygen and gets rid of carbon dioxide.

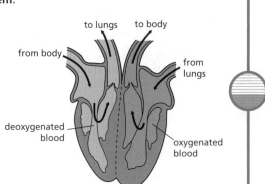

to lungs   to body

from body

from lungs

deoxygenated blood

oxygenated blood

The heart acts as a double pump.

- The three types of blood vessels are **arteries**, **capillaries** and **veins**.

Blood vessel	Transport of blood	Structure
Arteries	*Away from the heart* under high pressure	Walls are very thick, elastic and muscular to withstand the pressure.
Capillaries	Link arteries and veins	Walls are one cell thick to allow the transfer of substances to and from cells.
Veins	Collect blood and *return it to the heart*	Walls contain elastic, muscular tissue, but are thinner than those of arteries. The blood is under low pressure and veins have valves to prevent the backflow of blood.

- Heart muscle has its own blood supply. The **coronary arteries** run over the surface of the heart. They provide the heart with the nutrients and oxygen it needs to contract, and remove waste products.
- The coronary arteries can become blocked by fatty deposits. This will prevent the heart from receiving the oxygen it needs, leading to a heart attack.

Reducing the risk of heart disease

- **Coronary heart disease (CHD)** is caused by the build-up of fatty substances in the arteries.
- The main lifestyle factors that increase a person's risk of CHD are: **1** Smoking cigarettes; **2** Poor diet (a diet high in **saturated fat** and salt); **3** Misuse of drugs (this includes excessive consumption of alcohol); **4** Stress.

- Data show there is a link, or **correlation**, between these lifestyle factors and heart disease.
- Regular exercise helps to prevent against CHD by strengthening heart muscle, providing a healthy body weight and reducing stress.
- A healthy diet, low in saturated fat, lowers blood **cholesterol** and reduces the risk of heart disease.
- Genetic factors contribute to a person's chances of having CHD – members of families that have a history of heart disease need to lower the risk factors.

- Researchers study the occurrence of heart disease (epidemiological studies) and genetic factors.
- Rates of CHD are higher in industrialised countries, such as the UK and USA, than less-industrialised nations, for example India and China.

Ideas about science

You should be able to:

- use the ideas of correlation and cause when discussing data on a lifestyle factor, e.g. obesity, and heart disease
- explain why an observed correlation between heart disease and a lifestyle factor, such as poor diet, does not necessarily mean that the factor causes the outcome.

Improve your grade

The heart and circulatory system

Foundation: The heart needs its own blood supply to live. Describe how the heart receives this and what happens in a person with coronary heart disease.   *AO1* [5 marks]

Cardiovascular fitness

Heart rate and blood pressure

- **Heart rate** is measured by recording **pulse rate**. This is the number of pulses as blood passes through an artery close to the skin.

- Pulse rate is measured in beats per minute (bpm).

- The **resting heart rate** is the heart rate when a person is relaxed. A resting heart rate of 70–100 bpm is normal for teenagers. A rate of 50–70 bpm in an adult is an indication of a good level of fitness.

- The misuse of drugs, such as nicotine, alcohol and Ecstasy (MDMA), has a negative effect on health, including the heart rate, and increases the risk of heart disease and heart attack.

- **Blood pressure** measurements record the pressure of blood on the walls of an artery.

- People with consistently high blood pressure have an increased risk of heart disease.

- High blood pressure damages the walls of the arteries and makes them more likely to develop fatty deposits and get narrower. It also puts a strain on the heart.

- Blood needs to be under pressure to reach every cell in the body.

- Blood pressure is an important **indicator** of health.
  - **High blood pressure** increases the chance of strokes and heart attack.
  - **Low blood pressure** can cause dizziness and fainting.

- Blood pressure is measured as millimetres of mercury (mm Hg) and given as two numbers, for example 110/80. The higher value is when the heart is contracting; the lower value is when the heart is relaxed.

> **Remember!**
> There is no such thing as a single 'normal' value for heart and pulse rates and blood pressure. Everyone is different and values given are averages or ranges across the population.

Epidemiological studies

- Studies of the occurrence of disease using large numbers of individuals are called **epidemiological studies**.

- Epidemiological studies have been carried out on the link between lifestyle factors and heart disease. Studies are carried out:
  - on samples of individuals who are matched on as many factors as possible and differ only in the factor being investigated, e.g. smokers and non-smokers; drug users and non-drug users
  - on individuals chosen at random
  - that investigate whether the genes carried by individuals affect their risk of suffering from particular health problems.

The blood pressure measurement is a standard check that doctors carry out to check the general health of a patient.

Ideas about science

You should be able to:

- discuss whether given data suggests that a given factor, such as smoking, does or does not increase the chance of a given outcome, such as heart disease

- evaluate critically the design of a study to test if a given factor increases the chance of heart disease, by commenting on the design and results of the test.

Improve your grade

Heart rate and blood pressure

Foundation: Explain the term 'blood pressure' and describe how blood pressure affects health.

AO1 [2 marks]

Keeping things constant

A constant internal environment

- The maintenance of a constant environment is called **homeostasis**.

- Temperature, pH and levels of sugar, water and salt must be kept at values within a very narrow range for the body to function.

- Homeostasis involves communication by the nervous and hormonal systems.

- Response to change is by these 'automatic' control systems throughout the body.

- The systems involved in homeostasis are in three parts:

  1 Receptors detect change in the environment.

  2 Processing centres receive information and determine how the body will respond.

  3 Effectors produce a response.

Negative feedback is one way the body can restore its temperature if it rises or falls.

Water balance

- Water is *taken in* by drinking and eating and it's also *produced* by respiration.

- Water is lost in urine and faeces and when we sweat and breathe out.

- The water content of our bodies must be maintained to keep the body's cells bathed in **blood plasma**. So, the concentration of water in our cells must be kept constant.

- If the blood plasma is too concentrated, the cells will lose water.

- If the blood plasma is too dilute, the cells will absorb water and burst.

The kidneys

- Concentration of blood plasma is affected by external temperature, exercise level and intake of fluids and salt.

- The **kidneys** respond to changes in the blood plasma by changing the concentration of urine that is **excreted** from the body.

> **Remember!**
> You do not have to know how the kidneys work in detail. You just have to know that they play a vital role in balancing water, salt and other substances in the blood plasma.

- Some recreational drugs affect the water balance of the body:

  – Alcohol causes the kidneys to produce a large volume of dilute urine and the body becomes dehydrated.

  – Ecstasy (MDMA) causes the kidneys to produce very small volumes of concentrated urine. The body's cells will swell with water.

- The kidneys help to balance levels of water, **urea**, salts and other chemicals in the blood.

- **Anti-diuretic hormone (ADH)** is released by the pituitary gland in response to changes in the concentration of blood plasma. The secretion of ADH is controlled by negative feedback.

- ADH acts upon the kidneys to reduce the amount of water lost in the urine.

- Alcohol suppresses the release of ADH, so less water is reabsorbed by the kidneys.

- Ecstasy increases ADH production, so more water is reabsorbed by the kidneys.

> **EXAM TIP**
> Practise showing how ADH release is controlled by changes in the concentration of the blood plasma by drawing a flowchart.

Improve your grade

The kidneys

Higher: Explain the effect of Ecstasy on the water balance of a person's body.

AO1 [3 marks]

B2 Summary

Vaccines:
- contain a safe form of a disease-causing microorganism
- produce immunity (because memory cells remain after the vaccination)
- are very safe, but not risk-free (possible side effects; reactions vary because of genetic differences).

To prevent epidemics, a high percentage of the population must be vaccinated.

Organisms that cause infectious diseases include bacteria and viruses.

Damage to cells during an infection is because of toxins produced by the microorganisms.

Microorganisms reproduce very rapidly in the human body, to produce very large numbers.

Drug and vaccine trials ensure safety and effectiveness. The trials first involve animals and human cells. Later human trials involve healthy volunteers and people with the illness. Control groups use an existing drug or a placebo. The use of placebos raises ethical issues.

Trials can be:
- open-label (new drug is known by researchers and patient)
- 'blind' (the patients do not know who is receiving the new drug)
- 'double-blind' (neither patient nor researcher knows who is being given the new drug).

Protection against infection

The immune system has types of white blood cell that:
- destroy microorganisms by engulfing and digesting them
- produce antibodies against antigens on their surface.

Every antigen has a corresponding antibody that recognises it.

After an infection, memory cells remain in the body so that antibodies can be produced very quickly. The person now has immunity to the microorganism.

Antimicrobials:
- kill or inhibit the growth of bacteria, fungi and viruses
- include antibiotics, which are used to kill bacteria (only).

Over time, bacteria and fungi can develop resistance to antimicrobials.

To reduce this, only use antibiotics when necessary and complete the course.

Heart disease

The heart is part of the circulatory system. It is a double pump (left side to body, right side to lungs) and has its own blood supply.

The structure of the arteries, capillaries and veins is related to their functions.

Heart attacks are caused by fatty deposits blocking the blood supply to the heart.

Heart disease is caused by lifestyle factors (poor diet, stress, smoking, misuse of drugs), and/or genetic factors.

These factors are identified by large scale epidemiological and genetic studies.

Heart rate can be recorded by measuring the pulse rate.

Blood pressure is a measure of the pressure of the blood on an artery wall. It is measured as two numbers: the higher number is when the heart is contracting, the lower number when it is relaxed.

Values of 'normal' heart rate and blood pressure are given as a range, as they vary in individuals.

High blood pressure increases the risk of heart disease.

Water balance

Nervous and hormonal systems maintain a constant internal environment in the body (homeostasis). These control systems:
- are 'automatic'
- have receptors, processing centres and effectors.

Negative feedback is used to reverse any changes in the body's state.

Hormonal control of urine concentration is by ADH.

Alcohol reduces ADH secretion.

Ecstasy (MDMA) increases ADH secretion.

A balanced body water level maintains cell water concentration. This is vital for cell function.

Input of water is from drinks, food and respiration; losses are through sweating, breathing, faeces and urine.

The kidneys respond to water concentration in the plasma by producing dilute or concentrated urine.

Species adaptation, changes, chains of life

Species and adaptation

- A **species** is a group of organisms that can breed together to produce *fertile* offspring.
- Species are **adapted** to living in their environment. For example, a cactus is adapted to living in hot dry conditions by storing water in its stem; a camel stores fat in its hump.
- Adaptation of organisms to their environment means that they are able to *survive* to *reproduce* successfully.

- The organisms that live in a **habitat** are dependent on their environment and other species living there. They depend on other species for food and **compete** with each other for resources.
- Animals compete with each other for food, a mate, living space and territories. Plants compete for light, nutrients, water and space.
- The feeding relationships of organisms are shown in a **food web**.

- The feeding relationships of organisms in a habitat are often complex. They depend on each other, often in ways other than just providing food. This is called **interdependence**.
- Because of their interdependence, any change that affects one species in a food web is likely to affect all species in that food web.

Higher

Extinction

- A species can become **extinct** if it is unable to adapt rapidly to a change in the environment, e.g. climate change.
- Removal of habitats due to human activity threatens species, e.g. the Siberian tiger and mountain gorilla.

- The introduction of a new species can lead to extinction if the species is a competitor, predator or causes disease.

- The **extinction** of a species in a habitat will affect other organisms in the food web and may cause them to also become extinct.

Higher

Energy transfer

- Nearly all organisms on Earth are dependent on energy from the Sun.
- Plants absorb a small percentage of the energy from sunlight to produce their own food by **photosynthesis**. Plants store this energy in chemicals that make up the plants' cells and tissues.
- Other organisms get their energy by eating plants. Almost every food chain begins with a plant absorbing energy from the Sun.

- Energy is transferred from one organism to the next along a food chain.
- Only a small percentage of the energy transferred remains in each organism's body. In the transfer from a plant to an animal, some energy is lost because:
  - Some parts of the plant aren't eaten or can't be digested by the animal.
  - The animal uses some of the plant's energy for respiration. During respiration, some energy is lost as heat.
  - The waste products of the animal, for instance urine, contain some energy.
- Energy is *lost at each level of a food chain*. So the length of food chains is limited – they are rarely longer than four or five organisms.
- You can calculate the **efficiency** of energy transfer at any level using the equation:

$$\text{Percentage efficiency} = \frac{\text{energy in tissues}}{\text{energy in food eaten}} \times 100$$

- Energy transfer continues after an organism has died. Microorganisms such as bacteria and fungi feed on dead or decaying organisms. These are called **decomposers**.

- Partly decayed material is called detritus.
- **Detritivores**, such as earthworms and woodlice, feed on detritus and break it down further.

Higher

Improve your grade

Extinction

Foundation: The harlequin ladybird originated in Asia and arrived in Britain in 2004. It eats the same food as our native ladybird (aphids – greenfly and blackfly). It also **predates** on many insects (other ladybirds, and the eggs of butterflies and moths) and eats fruit.

Suggest why the spread of the harlequin ladybird might affect food webs.   *AO2* [4 marks]

Nutrient cycles, environmental indicators

The carbon cycle

- **Carbon** is the key element of the chemicals that make up all living things. It is continually recycled through the **carbon cycle**.

- Carbon *enters* the carbon cycle as carbon dioxide from the air. Plants *fix* this carbon, so that it can be used and stored by organisms, by photosynthesis.

- Carbon is *returned* to the air in the following ways:
  - as a product of **respiration**, when plants and animals release energy from food
  - through the **decomposition** of dead organisms by soil microorganisms such as bacteria and fungi
  - by **combustion** of organic materials.

The nitrogen cycle

- Nitrogen is an essential component of living things. It is recycled through the **nitrogen cycle**.

- Plants take up nitrogen from the soil through their roots, in the form of nitrogen compounds including **nitrates**. These are converted into **proteins**.

- Protein is an important nutrient in animals' diets. It passes along food chains as animals eat plants and other animals.

- Nitrates are released back into the soil as animals excrete waste, and as plants and animals die and are decomposed by microorganisms.

EXAM TIP

Make sure that you can identify the stages of nitrogen fixation and denitrification in a diagram of the nitrogen cycle.

- Nitrogen *enters* the nitrogen cycle in two ways:
  1 Nitrogen molecules in the air are split by lightning. Nitrogen atoms then combine with oxygen in the air to form nitrates, which are washed into the soil by rain.
  2 **Nitrogen-fixing bacteria**, found in the soil and in the roots of leguminous plants such as beans and peas, convert nitrogen in the air into nitrates.

- Nitrogen *leaves* the nitrogen cycle when **denitrifying bacteria** convert nitrates in the soil into nitrogen gas. This process is called denitrification.

Indicators of environmental change

- Environmental change can be measured using:
  - **non-living indicators**, e.g. carbon dioxide levels, temperature and nitrate levels
  - **living indicators**, e.g. **phytoplankton** (microscopic aquatic plant-like organisms), **lichens** (dual organisms made up of a fungus and alga living together) and aquatic organisms such as mayfly larvae.

- Scientists monitor environmental change from a local through to a global scale.

- Observations of some living indicators can give us very precise information about levels of pollution and environmental change. This is called a biotic index. For example, mayfly larvae need high levels of oxygen in the water, so will indicate very low levels of pollution.

- Interpretation of data from non-living indicators and living indicators helps scientists to monitor environmental change and trends over a period of time.

- Measurements using non-living indicators, e.g. CO_2 in the air and water, are monitored continuously. We can also look at historical levels from CO_2 trapped in ice.

Improve your grade

Indicators of environmental change

Higher: Lichens are sensitive to sulfur dioxide in the air. The distribution of three different types of lichen was measured in a city centre, and at different distances from it. The results are shown opposite.

  a Suggest what the graph tells you about the resistance of the lichens to pollution.
  AO3 [1 mark]

  b Why did the scientists measure the frequency of the lichens in 50 quadrats?
  AO2 [2 marks]

Variation and selection

Life on Earth

- Life on Earth began around 3500 million years ago. The first forms of life were very simple.
- Over millions of years these simple life forms gave rise to all the different species of organisms we see today, along with ones that are now extinct. This process of change is called **evolution**.
- The changes involved in evolution begin with variation between individuals. Variation has genetic and environmental causes.
- Evidence of how organisms changed over time is found in **fossils**. Fossils are the remains of organisms, or other traces of their lives such as footprints or eggs, that have turned into rock.
- Scientists can date fossils from the layer of rock they are found in.

Variation and evolution

- A **mutation** is a change in the genetic information in a cell. A mutation will result in a change in the characteristics of an organism.
- Mutations can occur as DNA is copied during the production of new cells.
- If a mutation occurs as sex cells are produced, the mutation is passed to the offspring.
- Most mutations are harmful, but sometimes new, useful characteristics are produced. Useful mutations will be passed on throughout the population.

- Evolution involves the development of new species. The theory of evolution provides the best explanation for the enormous number and variety of organisms on the planet today.
- In a population of organisms, the alleles of genes that occur in that population – the **gene pool** – will change because of mutations and other causes of genetic variation.
- It may be an advantage to have certain genes, rather than others, so that some become more common while others disappear. Over thousands or millions of years, with changes in the frequency of different genes, new species emerge.
- The theory of evolution is based on data and observations, both of organisms alive today and on the **fossil record**.

Variation and natural selection

- Because of genetic variation, some individuals will have characteristics that give them a better chance of survival than others.
- Individuals with advantageous genes will survive to reproduce and pass these to their offspring. This is **natural selection** – nature is selecting the most advantageous genes to be passed on.

- Selection does not just occur naturally – humans have produced crop plants, livestock and pets for thousands of years by a form of selection called **selective breeding**. This involves:
  - choosing the individuals with the characteristics that are closest to those required
  - breeding these (and preventing other individuals from breeding)
  - repeating the process over several generations.

- Natural selection is an important part of the evolutionary process. It results in an organism that is better able to survive in terms of:
  - reproduction, which will lead to an increase in the number of individuals displaying the characteristics in later generations.
  - competition with other animals, e.g. catching food, escaping predators, resistance to disease

Remember!
In selective breeding, humans are choosing the desirable characteristics. In natural selection it's nature that determines those individuals most able to survive.

Ideas about science

You should be able to:

- recognise data or observations (e.g. on the fossil record or the organisms on a volcanic island) that are accounted for by an explanation (such as evolution or natural selection)
- give good reasons for accepting or rejecting a proposed scientific explanation, such as evolution by natural selection.

Improve your grade

Variation and natural selection

Higher: Scientists have observed that the number of cases in which disease-causing bacteria are resistant to antibiotics has increased over the last 30 years. Explain why some people think that this is evidence of natural selection.

AO2 [4 marks]

Evolution, fossils and DNA

Natural selection and evolution

- Over a long period of time, advantageous genes chosen by natural selection are likely to become the norm in the population. This is how evolutionary change takes place.

- However, natural selection can be seen in operation over *short* periods of time. One example is the development of bacteria that have become resistant to antibiotics.

> **Remember!**
> It is natural selection, together with isolation or a changing environment, which provides the driving force for evolution.

- A number of factors influence the rate at which evolution takes place:
  - When the environment changes, only those organisms that are best adapted, or can re-adapt, will survive.
  - If organisms become **isolated**, for instance on an island, natural selection will act independently on the different populations. Over time the populations will become distinct and no longer able to reproduce with each other. They will be new species.

- In the search for evidence of evolution, scientists have investigated the relationships between organisms by:
  - examining the fossil record
  - observing similarities and differences in physical features, e.g. skeletons, flowers
  - analysing DNA sequences (more closely related organisms have more DNA sequences in common).

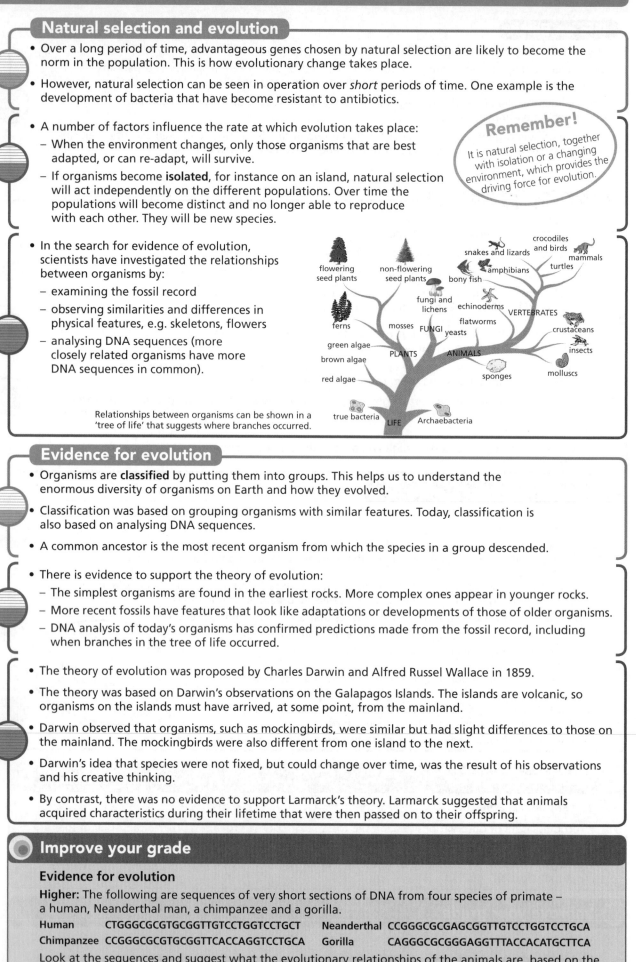

Relationships between organisms can be shown in a 'tree of life' that suggests where branches occurred.

Evidence for evolution

- Organisms are **classified** by putting them into groups. This helps us to understand the enormous diversity of organisms on Earth and how they evolved.

- Classification was based on grouping organisms with similar features. Today, classification is also based on analysing DNA sequences.

- A common ancestor is the most recent organism from which the species in a group descended.

- There is evidence to support the theory of evolution:
  - The simplest organisms are found in the earliest rocks. More complex ones appear in younger rocks.
  - More recent fossils have features that look like adaptations or developments of those of older organisms.
  - DNA analysis of today's organisms has confirmed predictions made from the fossil record, including when branches in the tree of life occurred.

- The theory of evolution was proposed by Charles Darwin and Alfred Russel Wallace in 1859.

- The theory was based on Darwin's observations on the Galapagos Islands. The islands are volcanic, so organisms on the islands must have arrived, at some point, from the mainland.

- Darwin observed that organisms, such as mockingbirds, were similar but had slight differences to those on the mainland. The mockingbirds were also different from one island to the next.

- Darwin's idea that species were not fixed, but could change over time, was the result of his observations and his creative thinking.

- By contrast, there was no evidence to support Larmarck's theory. Larmarck suggested that animals acquired characteristics during their lifetime that were then passed on to their offspring.

Improve your grade

Evidence for evolution

Higher: The following are sequences of very short sections of DNA from four species of primate – a human, Neanderthal man, a chimpanzee and a gorilla.

Human CTGGGCGCGTGCGGTTGTCCTGGTCCTGCT	**Neanderthal** CCGGGCGCGAGCGGTTGTCCTGGTCCTGCA
Chimpanzee CCGGGCGCGTGCGGTTCACCAGGTCCTGCA	**Gorilla** CAGGGCGCGGGGAGGTTTACCACATGCTTCA

Look at the sequences and suggest what the evolutionary relationships of the animals are, based on the data. Explain your reasoning. Explain the level of confidence you have in your conclusion, and suggest how this could be increased. *AO2, AO3* [5 marks]

Biodiversity and sustainability

Conserving biodiversity

- **Biodiversity** is the variety of life on Earth and in different habitats, including:
  - the number of different species
  - the range of different types of organisms, for example plants, animals, microorganisms
  - the **genetic diversity** (variation) within each species.
- Habitats such as tropical rainforests have a very high biodiversity (they have many species).
- Many of these species could be valuable to us as food crops or medicines.

- Species are now becoming extinct more rapidly than at any other time, except for **mass extinction events** seen in the fossil record. It is thought that this is connected with human activity, as organisms are hunted and their habitats are destroyed. Climate change will accelerate this rate of extinction.

- To record and monitor species accurately, they need to be classified.
- Organisms are classified in the following way:
  - A kingdom is a large group with many organisms but fewer characteristics in common.
  - Moving down from the kingdom, the groups get smaller and have fewer organisms with more characteristics in common.
  - The level identifying the individual type of organism is the species.

EXAM TIP

You do not need to know the names of the groups between a kingdom and a species, but in the exam you may be asked to sort organisms into different groups based on the number of features they have in common.

Sustainability

- **Sustainability** is about meeting today's needs without stopping future generations from meeting theirs, e.g. farming the land in a way that enables future generations to also farm it.
- Sustainability means: limiting our impact on wildlife, habitats and the environment; actively supporting ecosystems and populations of living organisms.

- To ensure sustainability, we need to maintain biodiversity. The loss of a single species removes a food supply and can have a big impact on the whole ecosystem.

- **Intensive monoculture crop production** maximises crop yields but it is not **sustainable**. It reduces the biodiversity of the field by: growing just one crop species; removing hedgerows to create huge fields for planting; spraying crops with herbicides and pesticides.

Improving sustainability

- Disposable products, e.g. nappies, create large amounts of waste that are slow to decompose.

- We can improve sustainability in product manufacture, for example by:
  - using as little energy as possible and minimal packaging
  - using locally available materials and limiting transport of the product
  - creating as little pollution as possible.
- The **Life Cycle Assessment** tracks the environmental impact of a product from:
  sourcing of raw materials → manufacture → transport → use → disposal.

- **Biodegradable** packaging has become popular for some products such as plastic carrier bags.
- It is best to reduce all types of packaging, as even biodegradable materials break down very slowly in landfill sites, produce carbon dioxide, and require energy to produce and transport.

Ideas about science

You should be able to:

- explain the idea of sustainability, and apply it to specific situations
- use Life Cycle Assessment data to compare the sustainability of alternative products or processes.

Improve your grade

Conserving biodiversity

Foundation: When investigating species extinctions, explain why we can use real data of human populations but need a computer model for numbers of extinctions. *AO2* [4 marks]

B3 Summary

Systems in balance

All organisms are dependent on energy from the Sun.

Plants absorb this energy. A small percentage is converted to chemical energy by photosynthesis, and stored in the chemicals that make up plant cells.

Energy transfer between organisms:

- occurs when they're eaten
- occurs after death, when their bodies are fed on by decomposers and detritivores
- involves only a small proportion of the energy – the majority is lost as heat, waste products and uneaten parts.

The length of food chains is therefore limited.

Environmental change can be monitored using non-living indicators and living indicators.

Nitrogen is recycled in the nitrogen cycle:

- Certain microorganisms convert nitrogen in the air into nitrates (nitrogen fixation).
- Nitrates are converted by plants to protein, and transferred to animals that eat them.
- Nitrogen compounds are returned to the soil by excretion, death and decay of organisms.
- Nitrogen is returned to the air by microorganisms (denitrification).

A species is a group of organisms that can breed together to produce fertile offspring.

Adaptation of a species to its environment increases its chances of survival.

Organisms in a habitat are dependent on the environment and other species for survival, and compete for resources.

In a food web, organisms show interdependence – changes affecting one species will impact on others.

A species can become extinct if the environment changes and it cannot adapt.

The carbon cycle:

- recycles carbon through the environment
- involves the processes of combustion, respiration, photosynthesis and decomposition.

Microorganisms are involved in respiration and decomposition.

Evolution

All individuals of a species show variation.

Variation that is genetic is passed on to offspring.

Variation can arise from accidental changes in genes (mutations).

Mutations in sex cells can be passed to offspring.

Life began on Earth 3500 million years ago.

All living things on Earth evolved from simple living things.

Natural selection results from genetic variation and competition between organisms for reproduction and survival.

Organisms that are better able to survive than others will pass on their genes.

Natural selection is similar to selective breeding by humans.

A combination of mutations, natural selection, environmental change and isolation produces new species by evolution.

Darwin's theory of evolution was the product of many observations and creative thought. Alternative theories, such as Lamarck's, do not fit with our understanding of genetics.

Evidence for evolution comes from the fossil record and analysis of DNA sequences of organisms.

Biodiversity

Organisms are classified into groups using physical features and DNA.

- A kingdom is the largest group of organisms.
- The term species identifies the particular type of organism.

Classifying organisms:

- makes sense of the enormous range of organisms on Earth
- helps to show their evolutionary relationships.

Biodiversity refers to the variety among living things (the number of species; the range of types of organisms; the genetic variation in a species).

The rate of species extinction is increasing as a result of human activity.

Biodiversity is important for the future growth of crops and search for new medicines.

Sustainability is meeting the needs of the people today without damaging the Earth for future generations.

Biodiversity is essential for sustainability. In crop cultivation, large scale monoculture is not sustainable.

Improvements in sustainable production include the selection of packaging materials.

The chemical reactions of living things

Photosynthesis and respiration

- **Photosynthesis** is the process by which *green plants* make their own food. During the process, light energy is converted to chemical energy. The end product of photosynthesis is **glucose**.

- Photosynthesis plays a vital role in making energy available to organisms through food chains.

- **Respiration** is the process by which *all organisms* release energy from food.

- At night, plants respire only and give out **carbon dioxide**. During the day, plants photosynthesise as well as respire. There will be a net output of oxygen.

Remember!
All organisms respire; only plants photosynthesise.

- Organisms use energy for many activities, including movement and producing large **molecules** required for growth.

Enzymes

- **Enzymes** are chemicals that speed up the rate of chemical reactions.

- Enzymes are **proteins**. They consist of long chains of **amino acids** joined together.

- Cells assemble enzymes using the instructions provided in genes.

- The chemicals that enzymes work on are called **substrates**.

- The chemicals produced in the reaction are called **products**.

Remember!
Enzymes are *destroyed* by high temperatures. They are just chemicals, so they can't be 'killed'!

- The order and types of amino acid in an enzyme give it a complicated three-dimensional shape. This shape is essential for the enzyme to work.

- Part of the enzyme called the **active site** has a special shape that the substrate fits neatly into, like a key in a lock.

- Substrate molecules locked into the active site take part in the chemical reaction, and product molecule(s) are released. This is the lock and key model of enzyme action.

- Enzymes need a specific **pH** and temperature to work at their optimum. They stop working if the pH is inappropriate or the temperature is too high.

- The body temperature of mammals and birds is around 37 °C. This is the optimum temperature of *most* enzymes.

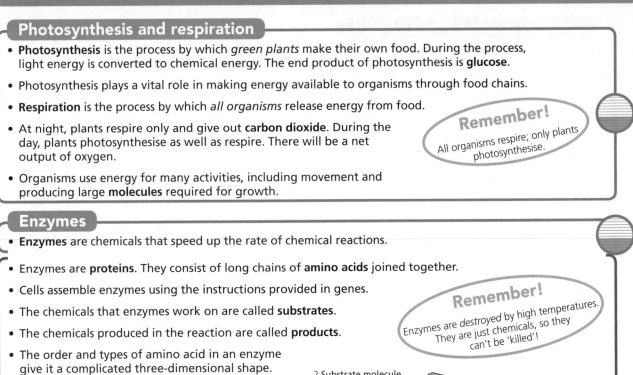

2 Substrate molecule fits in active site in enzyme.

3 Reaction occurs and products made; the enzyme speeds up this reaction.

1 Substrate molecules move towards active site in enzyme.

4 Product molecules do not fit in the enzymes as well so are released. The enzyme can then be used again with new substrate molecules.

The lock and key model of enzyme action.

- As the temperature is increased, enzyme activity increases because the reaction rate increases.

- High temperatures change the shape of the active site of the enzyme. At a point where the change in shape is permanent, the enzyme has been **denatured**.

- pH also affects enzyme activity by changing the shape of the active site. The change in shape can be temporary or permanent (it is denatured).

EXAM TIP

In the exam it is likely that you will be asked to interpret information on the activity of enzymes from data or graphs. Practise answering this type of question.

Higher

Ideas about science

You should be able to:
- in experiments on enzymes:
  - when planning an investigation, identify the effect of factors on the outcome, and control factors that might affect the outcome, other than the one being investigated
  - calculate the mean of a set of repeated measurements
  - when asked to evaluate data, make reference to its repeatability
  - use data (rather than opinion) to justify an explanation.

Improve your grade

Enzymes
Higher: Explain what happens to enzyme activity as the temperature is increased.   *AO1* [5 marks]

How do plants make food?

Glucose: making it and using it

- Photosynthesis is a series of chemical reactions that use energy from sunlight to build large food molecules in plant cells and some microorganisms such as **phytoplankton**.

- It is summarised in the word equation: carbon dioxide + water $\xrightarrow{\text{light energy}}$ glucose + oxygen

- Light energy from the Sun is required to drive the reaction between carbon dioxide and water to build up glucose.

- Sunlight is absorbed by the green chemical **chlorophyll** which, along with the enzymes for photosynthesis, is found in structures in the cell called **chloroplasts**.

- Chloroplasts are found in parts of the plant exposed to sunlight.

- The product of photosynthesis, glucose, is:
  - converted to chemicals needed for the plant's growth, e.g. **cellulose**, protein, chlorophyll
  - converted into starch for storage
  - used in respiration to release energy.

- Glucose from photosynthesis and **nitrates** taken up by the plant roots are used to synthesise amino acids, which are assembled into proteins.

- Oxygen is produced as a waste product of photosynthesis.

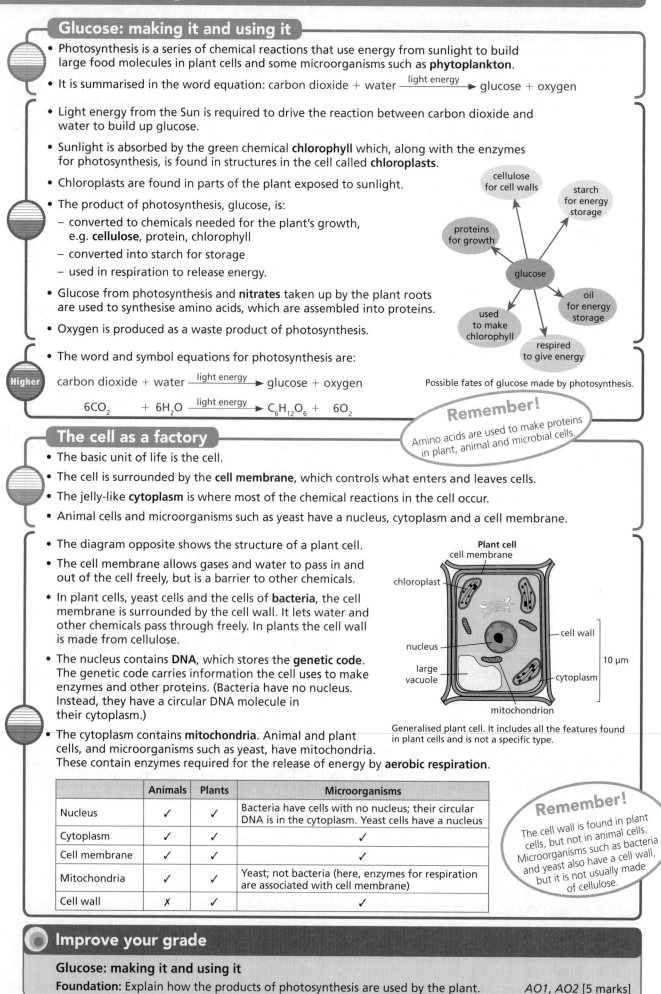

Possible fates of glucose made by photosynthesis.

- The word and symbol equations for photosynthesis are:

Higher

carbon dioxide + water $\xrightarrow{\text{light energy}}$ glucose + oxygen

$$6CO_2 + 6H_2O \xrightarrow{\text{light energy}} C_6H_{12}O_6 + 6O_2$$

Remember!
Amino acids are used to make proteins in plant, animal and microbial cells.

The cell as a factory

- The basic unit of life is the cell.
- The cell is surrounded by the **cell membrane**, which controls what enters and leaves cells.
- The jelly-like **cytoplasm** is where most of the chemical reactions in the cell occur.
- Animal cells and microorganisms such as yeast have a nucleus, cytoplasm and a cell membrane.

- The diagram opposite shows the structure of a plant cell.
- The cell membrane allows gases and water to pass in and out of the cell freely, but is a barrier to other chemicals.
- In plant cells, yeast cells and the cells of **bacteria**, the cell membrane is surrounded by the cell wall. It lets water and other chemicals pass through freely. In plants the cell wall is made from cellulose.
- The nucleus contains **DNA**, which stores the **genetic code**. The genetic code carries information the cell uses to make enzymes and other proteins. (Bacteria have no nucleus. Instead, they have a circular DNA molecule in their cytoplasm.)
- The cytoplasm contains **mitochondria**. Animal and plant cells, and microorganisms such as yeast, have mitochondria. These contain enzymes required for the release of energy by **aerobic respiration**.

Plant cell

Generalised plant cell. It includes all the features found in plant cells and is not a specific type.

	Animals	Plants	Microorganisms
Nucleus	✓	✓	Bacteria have cells with no nucleus; their circular DNA is in the cytoplasm. Yeast cells have a nucleus
Cytoplasm	✓	✓	✓
Cell membrane	✓	✓	✓
Mitochondria	✓	✓	Yeast; not bacteria (here, enzymes for respiration are associated with cell membrane)
Cell wall	✗	✓	✓

Remember!
The cell wall is found in plant cells, but not in animal cells. Microorganisms such as bacteria and yeast also have a cell wall, but it is not usually made of cellulose.

Improve your grade

Glucose: making it and using it
Foundation: Explain how the products of photosynthesis are used by the plant.

AO1, AO2 [5 marks]

Providing the conditions for photosynthesis

Moving chemicals in and out of plants by diffusion

- **Diffusion** is the movement of chemicals from a high to low concentration.

- In photosynthesis:
  – Water is taken up by plant roots.
  – Carbon dioxide enters plant leaves by diffusion.
  – Oxygen leaves plant leaves by diffusion.

- Diffusion is:
  – passive (it just happens, and does not require energy)
  – the movement of molecules from an area where they are in high concentration, to an area where they are in lower concentration.

Remember!
Diffusion happens because of the random movement of molecules. This movement does not stop when the concentrations of molecules in two areas are equal; there is just no *overall* movement.

Moving chemicals in and out of plants by osmosis

- **Osmosis** is a special kind of diffusion involving water.

- It happens when chemicals are separated by a **partially permeable membrane**, e.g. the cell membrane. The cell membrane lets water pass through, but keeps other chemicals in or out.

- Osmosis is the overall movement of water from an area of high concentration (water, or a dilute solution) to an area of low concentration.

- The movement of water into plant roots from the soil, and across the roots, occurs by osmosis.

EXAM TIP

You may be asked to plot data showing the movement of water into and out of plant tissue and explain this movement, draw conclusions about the concentration of cell sap, and evaluate the data.

Active transport

- Minerals taken up by plant roots are used to make chemicals essential to cells, e.g. nitrogen taken up as nitrates is used to make proteins.

- Nitrates are normally in a higher concentration in plant cells than the soil, so root cells cannot take up nitrates by diffusion.

- Root cells use **active transport** to take up nitrates.

- Active transport uses energy from respiration to transport chemicals across cell membranes.

Higher

Factors that limit the rate of photosynthesis

- The rate of photosynthesis can be limited by temperature, carbon dioxide and light intensity.

- Carbon dioxide is present in the air at only 0.04%, so commercial plant growers often increase levels in their greenhouses.

- If a factor such as light intensity or carbon dioxide concentration is increased, the rate of photosynthesis increases, and then levels off.

- At the point where the graph levels off, something is limiting the rate. This is a **limiting factor**.

- A rise in temperature increases the rate of photosynthesis, up to a certain point. The rate then decreases, because of the effect of temperature on enzyme activity.

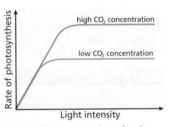

Increasing the concentration of carbon dioxide increases the rate of photosynthesis until another factor becomes limiting.

Here, light is limiting, so the rate can be increased by increasing the light intensity.

Higher

Ideas about science

You should be able to:

- in an investigation on osmosis: outline how a proposed scientific explanation might be tested; identify and treat **outliers** in a set of data

- in an investigation on the rate of photosynthesis: identify the outcome and factors that may affect it; suggest how an outcome might alter when a factor is changed.

Improve your grade

Moving chemicals in and out of plants by osmosis

Higher: Explain how plant roots take up water, and how this water moves across a plant root.

AO2 [5 marks]

Fieldwork to investigate plant growth

Investigating the effects of light on plant growth

- When investigating the effects of light intensity on plant growth, whether at different locations within an area or in different areas, ecologists need to:
  - Use an **identification key**, such as the one shown opposite, to identify the plants they find.
  - Measure how common the plants are.
  - Use a light meter to measure the light intensity.

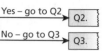

- Ecologists investigate the **abundance** and distribution of plants:
  - at different locations in an area, and make a comparison
  - in different areas, and make a comparison.

- It is usually not appropriate to count all the plants in the area, so ecologists take appropriate samples.

- A sample is usually taken using a metal or wooden frame of known area called a **quadrat**.

- The ecologist makes a decision about the size of quadrat to use based on the size of the organism, and the size of area to be sampled.

- A quadrat is put on the ground:
  - a number of times, e.g. 10, so that an average is taken, and the number per m² can be calculated
  - at random in a location (by throwing, or preferably, picking coordinates at random from a grid system), *if there is no obvious change in the distribution of plants in the area.*

Quadrats used to survey an area.

> **Remember!**
> By using information of plant distribution from quadrats, ecologists can relate the plants' distribution to the availability of light and other factors.

- Alternatively, *if there is an obvious change in the plants in the area*, use the quadrat at measured distances along a line called a **transect**.

- Trees, bushes and hedges, or a group of the same species of plant, can cast shade on an area and therefore affect the growth of plants around them.

- There may therefore be a **correlation** between the distance from a tree, bush and hedge and the growth of other plants, and the growth of plants grown at high and low densities.

- However, other factors may be involved, such as competition for nutrients.

> **EXAM TIP**
> You may be asked to analyse, interpret and evaluate data on the effect of light, as well as other factors, on the growth of plants in an area.

Ideas about science

You should be able to:
- for data of plant distribution:
  - identify where a correlation exists
  - use the ideas of correlation and cause when discussing data and show awareness that a correlation does not necessarily indicate a causal link
  - evaluate critically the design of a field study, e.g. commenting on sample size and how well the samples are matched.

Improve your grade

Investigating the effects of light on plant growth

Foundation: Describe and explain how an ecologist would compare how a plant is distributed in two meadows.

AO1, AO2 [5 marks]

How do living things obtain energy?

Aerobic respiration

- Life processes depend on energy released from food by **respiration**. Respiration is a series of chemical reactions in cells that release energy by breaking down large food molecules.

- All organisms respire, including animals, plants and microorganisms.

- Respiration occurs in every cell in the body.

- The chemical used for respiration is glucose (chemicals in our food are converted into glucose, or we eat glucose itself).

- Respiration using oxygen is called **aerobic respiration**. The word equation for this is:

  glucose + oxygen ⟶ carbon dioxide + water (+ energy released)

- Aerobic respiration takes place in animal and plant cells, and some microorganisms.

- Organisms require the energy released by respiration for the synthesis of large molecules and movement.

- Plants use the food produced by photosynthesis for respiration and **active transport**. Excess glucose is stored as starch.

- Respiration takes place as a series of enzyme-controlled reactions, with energy being released in stages.

> **Remember!**
> Don't forget that plants carry out respiration as well as photosynthesis. It's just that during bright daylight, photosynthesis occurs much faster than respiration, so the plant gives out oxygen and not carbon dioxide.

Respiration   small amounts of energy given out at each step

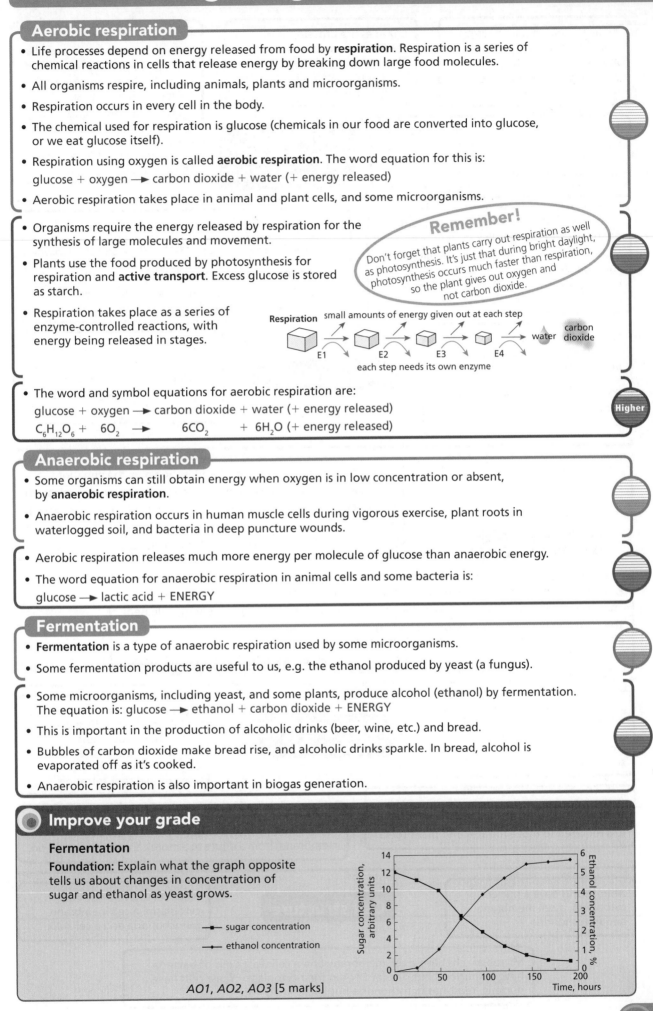

E1   E2   E3   E4   water   carbon dioxide

each step needs its own enzyme

- The word and symbol equations for aerobic respiration are:

  glucose + oxygen ⟶ carbon dioxide + water (+ energy released)

  $C_6H_{12}O_6 + 6O_2 \longrightarrow 6CO_2 + 6H_2O$ (+ energy released)

Higher

Anaerobic respiration

- Some organisms can still obtain energy when oxygen is in low concentration or absent, by **anaerobic respiration**.

- Anaerobic respiration occurs in human muscle cells during vigorous exercise, plant roots in waterlogged soil, and bacteria in deep puncture wounds.

- Aerobic respiration releases much more energy per molecule of glucose than anaerobic energy.

- The word equation for anaerobic respiration in animal cells and some bacteria is:

  glucose ⟶ lactic acid + ENERGY

Fermentation

- **Fermentation** is a type of anaerobic respiration used by some microorganisms.

- Some fermentation products are useful to us, e.g. the ethanol produced by yeast (a fungus).

- Some microorganisms, including yeast, and some plants, produce alcohol (ethanol) by fermentation. The equation is: glucose ⟶ ethanol + carbon dioxide + ENERGY

- This is important in the production of alcoholic drinks (beer, wine, etc.) and bread.

- Bubbles of carbon dioxide make bread rise, and alcoholic drinks sparkle. In bread, alcohol is evaporated off as it's cooked.

- Anaerobic respiration is also important in biogas generation.

Improve your grade

Fermentation

Foundation: Explain what the graph opposite tells us about changes in concentration of sugar and ethanol as yeast grows.

—■— sugar concentration

—+— ethanol concentration

AO1, AO2, AO3 [5 marks]

B4 Summary

All living cells require energy from respiration.

Photosynthesis in plants makes food and energy available to food chains.

Enzymes are proteins that speed up chemical reactions. They are assembled in the cytoplasm from the instructions carried by genes.

Each enzyme works best at an optimum temperature and pH.

The rate of an enzyme-controlled reaction increases as temperature increases.

The chemical(s) that an enzyme works on is the substrate. This must be the correct shape to fit to the active site of the enzyme – the lock and key mechanism.

Chemical reactions in cells

High temperatures and extremes of pH prevent enzyme action by changing the shape of the active site (denaturing the enzyme).

Photosynthesis uses energy from sunlight to build large food molecules.

Light is absorbed by the green chemical, chlorophyll.

Chloroplasts contain chlorophyll and the enzymes needed for photosynthesis.

The equation for photosynthesis is:

$$\text{carbon dioxide} + \text{water} \xrightarrow{\text{light energy}} \text{glucose} + \text{oxygen}$$

$$6CO_2 + 6H_2O \xrightarrow{\text{light energy}} C_6H_{12}O_6 + 6O_2$$

The glucose produced is converted to other chemicals that the plant needs, e.g. starch, cellulose and protein.

Fieldwork techniques are used to investigate the effect of light on plants in the wild. Identification keys show the types of plants affected. Plant distribution is shown using quadrats placed at random or along a transect.

Photosynthesis

The rate of photosynthesis is:
- affected by temperature, carbon dioxide and light intensity
- limited if any one of these is in short supply.

Respiration releases the energy from food used to drive chemical reactions in cells.

Aerobic respiration uses oxygen, and is shown by the equation:

$$\text{glucose} + \text{oxygen} \longrightarrow \text{carbon dioxide} + \text{water} \text{ (+ energy released)}$$

$$C_6H_{12}O_6 + 6O_2 \longrightarrow 6CO_2 + 6H_2O \text{ (+ energy released)}$$

It is carried out by animal cells, plant cells and some microorganisms.

A series of enzyme-controlled reactions release energy in stages.

Respiration

Anaerobic respiration takes place when oxygen is absent (or in very low concentration).

The equation in animal cells and some bacteria is: glucose $\longrightarrow$ lactic acid (+ energy released)

The equation in plant cells and some microorganisms including yeast is:
glucose $\longrightarrow$ ethanol + carbon dioxide (+ energy released)

It is important economically in the production of alcoholic drinks, bread, yogurt and biogas.

The cell membrane regulates what enters and leaves cells. It allows gases and water to move freely but is a barrier to other substances.

Diffusion and osmosis

O_2 and CO_2 move into and out of leaves by diffusion – the passive movement of molecules from an area of high concentration to an area of low concentration.

Active transport of molecules uses energy to transport molecules across the cell membrane.

It is required to move nitrates into plant roots.

Osmosis (a special form of diffusion) is the overall movement of water, through a partially permeable membrane, from a dilute to a more concentrated solution.

Animal cells have a nucleus, cytoplasm, a cell membrane and mitochondria.

Cell structure

Plant cells have a nucleus, cytoplasm, a cell membrane, mitochondria and a cell wall.

Microbial cells have a cell membrane and a cell wall. Cells of bacteria have no nucleus, but circular DNA in their cytoplasm. Yeast cells have a nucleus. Yeast has mitochondria but bacteria do not.

How organisms develop

Cell specialisation in animals

- In organisms that are **multicellular**, cells are **specialised** to do different jobs.

- Cells of the same type are grouped into **tissues**, e.g. muscle cells → muscular tissue.

- Different tissues are grouped together, and work together, in **organs**. For example, the heart has muscular tissue, epithelial (lining and covering) tissue, blood and nervous tissue.

- Organs work together as body systems, e.g. the circulatory system.

- Organisms begin life as a **zygote** – a fertilised egg.

- The zygote divides by **mitosis** – into 2, 4, 8, etc. – to form an **embryo**.

- In humans, up to and including the eight-cell stage, the cells are *identical*. These cells are **embryonic stem cells** – they will produce *any* cell type in the body.

- After the eight-cell stage, cells become specialised (this is called **differentiation**), and different tissues form.

- As adults, stem cells remain in certain parts of the body. **Adult stem cells** can differentiate into a limited number of cell types, e.g. bone marrow cells into different types of blood cell.

- In specialised cells, only the **genes** are needed to enable the cell to function, as that type of cell is switched on. In embryonic stem cells, any gene can be switched on.

Remember!
All body cells have the same genes. In most cells, only those required for the cells to function are switched on.

Higher

Cell specialisation in plants

- Specialised plant cells form tissues such as the **xylem**, which transports water and mineral salts, and **phloem**, which transports the products of photosynthesis.

- Tissues are organised into organs, e.g. stems, leaves, roots and flowers.

- Cells in regions called **meristems** are unspecialised.

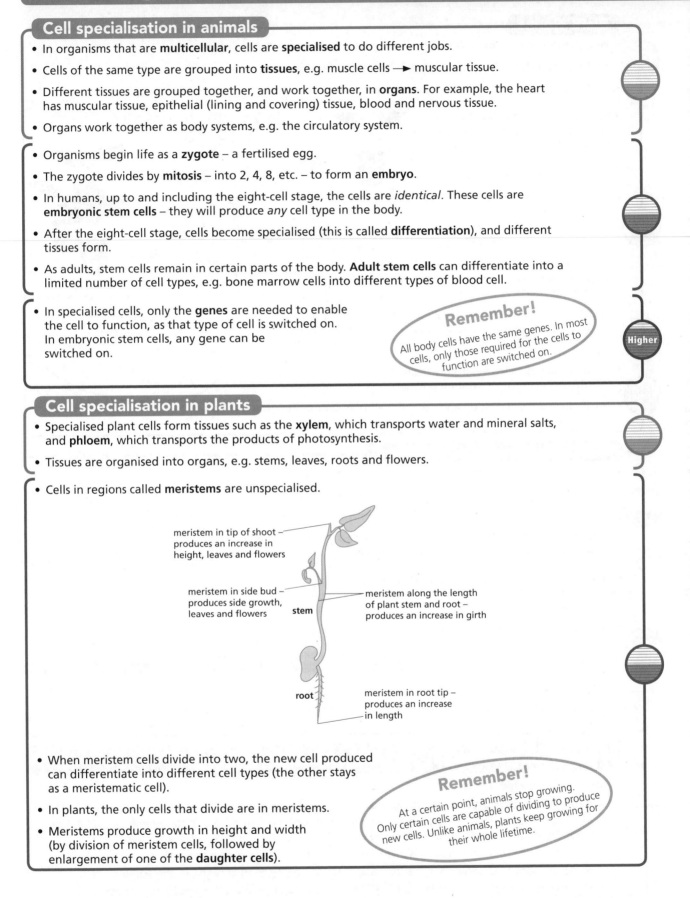

meristem in tip of shoot – produces an increase in height, leaves and flowers

meristem in side bud – produces side growth, leaves and flowers

stem

meristem along the length of plant stem and root – produces an increase in girth

root

meristem in root tip – produces an increase in length

- When meristem cells divide into two, the new cell produced can differentiate into different cell types (the other stays as a meristematic cell).

- In plants, the only cells that divide are in meristems.

- Meristems produce growth in height and width (by division of meristem cells, followed by enlargement of one of the **daughter cells**).

Remember!
At a certain point, animals stop growing. Only certain cells are capable of dividing to produce new cells. Unlike animals, plants keep growing for their whole lifetime.

⊙ Improve your grade

Cell specialisation in animals
Foundation: Compare how cells become specialised in animals and plants.

AO2 [5 marks]

Plant development

Plant clones

- New plants can be grown by placing the cut end of a shoot in water or soil.
- Roots grow at the base of the stem, while the shoot continues to grow.
- Plants grown in this way include garden plants and houseplants.

- Pieces of plants, e.g. plant stems, that have meristems and are used to produce **clones**, are called cuttings.
- Cuttings:
  - can be used to produce new plants with the same desirable features as the parent
  - produce clones that are genetically identical to the parent plant.
- Root growth in cuttings is promoted by plant **hormones** (using hormone rooting powder).
- Another method of cloning is called **tissue culture** – a small piece of tissue, or a few cells are placed on agar jelly containing nutrients and plant hormones. Each will grow into a small plant or plantlet.
- Plant hormones called **auxins** are included in the agar for tissue culture and in hormone rooting powder.
- **Higher** Auxins increase cell division and cell enlargement, promoting growth of the plant tissue.

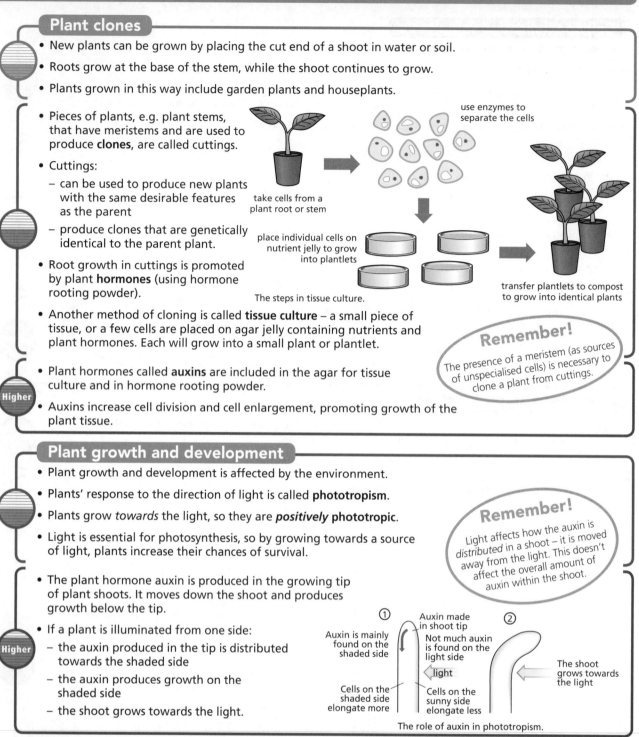

take cells from a plant root or stem

use enzymes to separate the cells

place individual cells on nutrient jelly to grow into plantlets

transfer plantlets to compost to grow into identical plants

The steps in tissue culture.

Remember!
The presence of a meristem (as sources of unspecialised cells) is necessary to clone a plant from cuttings.

Plant growth and development

- Plant growth and development is affected by the environment.
- Plants' response to the direction of light is called **phototropism**.
- Plants grow *towards* the light, so they are *positively* phototropic.
- Light is essential for photosynthesis, so by growing towards a source of light, plants increase their chances of survival.

- **Higher** The plant hormone auxin is produced in the growing tip of plant shoots. It moves down the shoot and produces growth below the tip.
- If a plant is illuminated from one side:
  - the auxin produced in the tip is distributed towards the shaded side
  - the auxin produces growth on the shaded side
  - the shoot grows towards the light.

Remember!
Light affects how the auxin is *distributed* in a shoot – it is moved away from the light. This doesn't affect the overall amount of auxin within the shoot.

① Auxin made in shoot tip
Auxin is mainly found on the shaded side
Not much auxin is found on the light side
light
② The shoot grows towards the light
Cells on the shaded side elongate more
Cells on the sunny side elongate less

The role of auxin in phototropism.

Ideas about science

You should be able to:
- in experiments on phototropism and auxins:
  - in an account of scientific work, identify statements which report data and statements of explanatory ideas (hypotheses, explanations, theories)
  - identify where creative thinking is involved in the development of an explanation
  - recognise data or observations that are accounted for by, or conflict with, an explanation
  - draw valid conclusions about the implications of given data for a given scientific explanation.

Improve your grade

Plant clones

Foundation: Explain how and why plant breeders who have produced a new variety of plant take many cuttings from it.

AO1, AO2 [5 marks]

Student Book pages 46–47 and 50–51

Cell division

Mitosis

- **Mitosis** is the type of cell division that takes place when an organism grows, and cells divide to repair tissues.

- Mitosis results in the production of *two* **daughter cells** that are genetically identical, i.e. have the same number of **chromosomes** as the parent cell.

- Before mitosis, the **DNA** in each chromosome is copied. Each chromosome is now a double chromosome, with two DNA molecules.

- During mitosis, each double chromosome separates, so that two nuclei and two cells are produced.

- The events between and leading up to cell division, and cell division itself, are called the cell cycle. The main processes of the cell cycle are:

  1 *Cell growth:* the cell increases in size; numbers of organelles increase; the DNA in each chromosome is copied.

  2 *Mitosis:* two daughter cells, each identical to the parent cell and containing an identical set of chromosomes, are produced as the strands of each double chromosome separate and two nuclei are formed.

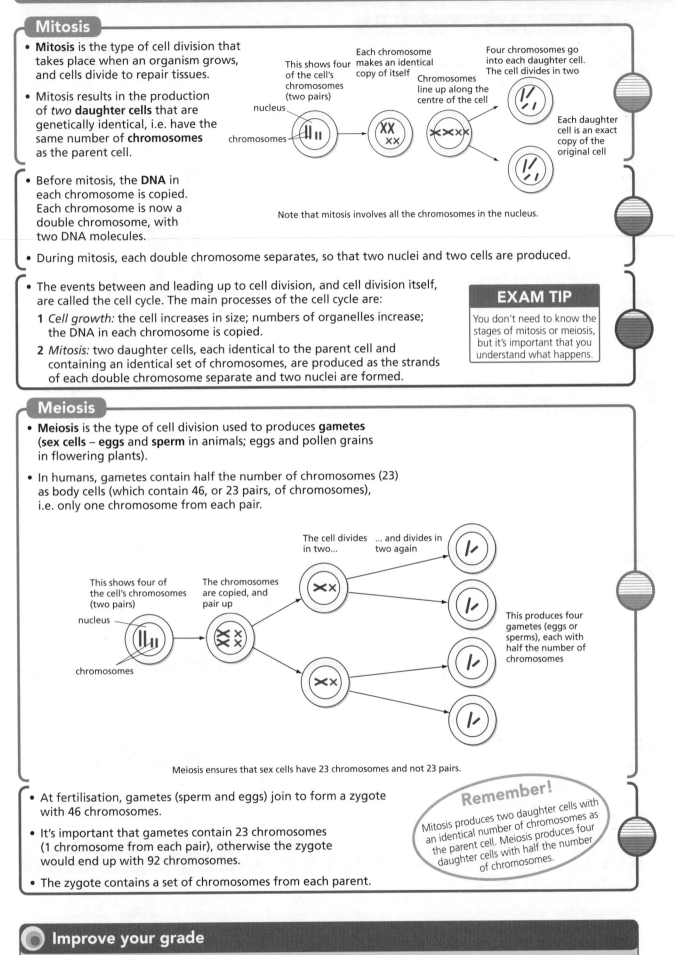

This shows four of the cell's chromosomes (two pairs)

nucleus

chromosomes

Each chromosome makes an identical copy of itself

Chromosomes line up along the centre of the cell

Four chromosomes go into each daughter cell. The cell divides in two

Each daughter cell is an exact copy of the original cell

Note that mitosis involves all the chromosomes in the nucleus.

> **EXAM TIP**
>
> You don't need to know the stages of mitosis or meiosis, but it's important that you understand what happens.

Meiosis

- **Meiosis** is the type of cell division used to produces **gametes** (**sex cells** – **eggs** and **sperm** in animals; eggs and pollen grains in flowering plants).

- In humans, gametes contain half the number of chromosomes (23) as body cells (which contain 46, or 23 pairs, of chromosomes), i.e. only one chromosome from each pair.

This shows four of the cell's chromosomes (two pairs)

nucleus

chromosomes

The chromosomes are copied, and pair up

The cell divides in two...

... and divides in two again

This produces four gametes (eggs or sperms), each with half the number of chromosomes

Meiosis ensures that sex cells have 23 chromosomes and not 23 pairs.

- At fertilisation, gametes (sperm and eggs) join to form a zygote with 46 chromosomes.

- It's important that gametes contain 23 chromosomes (1 chromosome from each pair), otherwise the zygote would end up with 92 chromosomes.

- The zygote contains a set of chromosomes from each parent.

> **Remember!**
>
> Mitosis produces two daughter cells with an identical number of chromosomes as the parent cell. Meiosis produces four daughter cells with half the number of chromosomes.

Improve your grade

Mitosis and meiosis

Higher: Explain why gametes (sex cells) are produced by meiosis and not by mitosis.   *AO1, AO2* [4 marks]

Chromosomes, genes, DNA and proteins

Chromosomes, genes and DNA

- Chromosomes:
  - are thread-like structures found in the nucleus
  - are made from a DNA molecule
  - can be grouped into pairs (humans have 23 pairs).

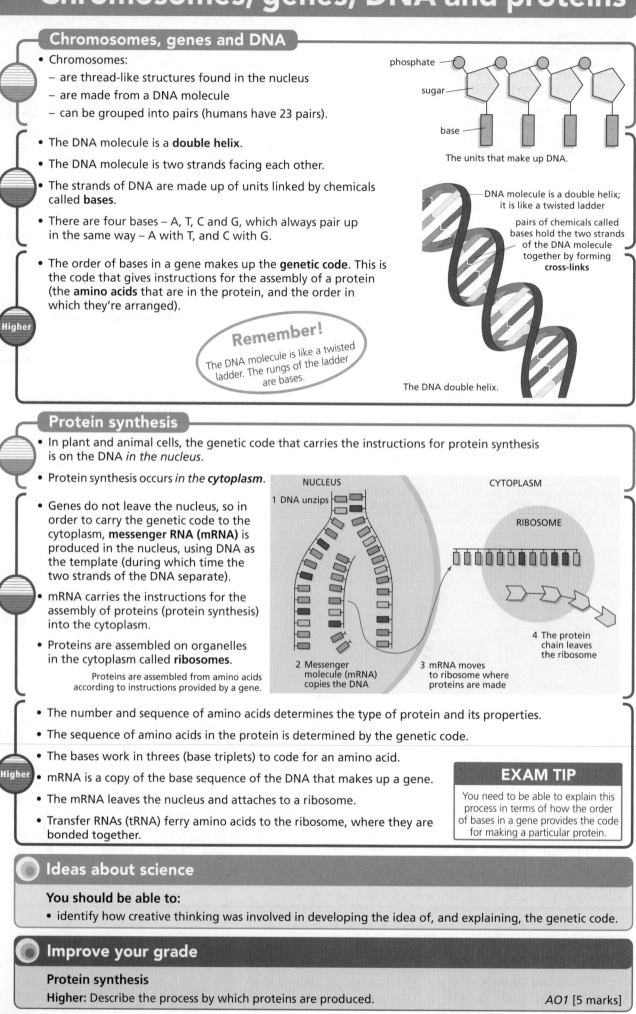

The units that make up DNA.

- The DNA molecule is a **double helix**.

- The DNA molecule is two strands facing each other.

- The strands of DNA are made up of units linked by chemicals called **bases**.

- There are four bases – A, T, C and G, which always pair up in the same way – A with T, and C with G.

- The order of bases in a gene makes up the **genetic code**. This is the code that gives instructions for the assembly of a protein (the **amino acids** that are in the protein, and the order in which they're arranged).

Higher

DNA molecule is a double helix; it is like a twisted ladder

pairs of chemicals called bases hold the two strands of the DNA molecule together by forming **cross-links**

Remember!
The DNA molecule is like a twisted ladder. The rungs of the ladder are bases.

The DNA double helix.

Protein synthesis

- In plant and animal cells, the genetic code that carries the instructions for protein synthesis is on the DNA *in the nucleus*.

- Protein synthesis occurs *in the **cytoplasm***.

- Genes do not leave the nucleus, so in order to carry the genetic code to the cytoplasm, **messenger RNA (mRNA)** is produced in the nucleus, using DNA as the template (during which time the two strands of the DNA separate).

- mRNA carries the instructions for the assembly of proteins (protein synthesis) into the cytoplasm.

- Proteins are assembled on organelles in the cytoplasm called **ribosomes**.

Proteins are assembled from amino acids according to instructions provided by a gene.

NUCLEUS
1 DNA unzips
2 Messenger molecule (mRNA) copies the DNA

CYTOPLASM
RIBOSOME
3 mRNA moves to ribosome where proteins are made
4 The protein chain leaves the ribosome

- The number and sequence of amino acids determines the type of protein and its properties.

- The sequence of amino acids in the protein is determined by the genetic code.

- The bases work in threes (base triplets) to code for an amino acid.

Higher
- mRNA is a copy of the base sequence of the DNA that makes up a gene.

- The mRNA leaves the nucleus and attaches to a ribosome.

- Transfer RNAs (tRNA) ferry amino acids to the ribosome, where they are bonded together.

EXAM TIP

You need to be able to explain this process in terms of how the order of bases in a gene provides the code for making a particular protein.

Ideas about science

You should be able to:
- identify how creative thinking was involved in developing the idea of, and explaining, the genetic code.

Improve your grade

Protein synthesis
Higher: Describe the process by which proteins are produced.

AO1 [5 marks]

Cell specialisation

Switching genes on and off

- The cell only produces the proteins it needs to carry out its function.
- The **genes** to make these proteins are **switched on**; the others are **switched off**.

- Up to the eight-cell stage of the embryo, the cells (embryonic stem cells) are identical.
- The cells produced by the division of embryonic stem cells undergo differentiation to produce specialised cells.
- Specialised cells begin to make specific proteins. They usually change shape and structure, e.g. muscle cells produce the proteins that enable them to contract.
- In embryonic stem cells, any gene can be switched on, so they can produce any type of cell.
- Embryonic stem cells (and adult stem cells) therefore have the potential to replace cells needed to replace damaged tissues.

Stem cell research and therapy

- Stem cells are used to produce new cells to replace damaged or diseased cells.
- Adult stem cells are found at various locations in the body, e.g. the bone marrow.
- These cells can be used to produce a limited number of cell types, e.g. bone marrow cells will differentiate to produce types of blood cells.

> **EXAM TIP**
>
> Make sure that you understand the difference between embryonic and adult stem cells.

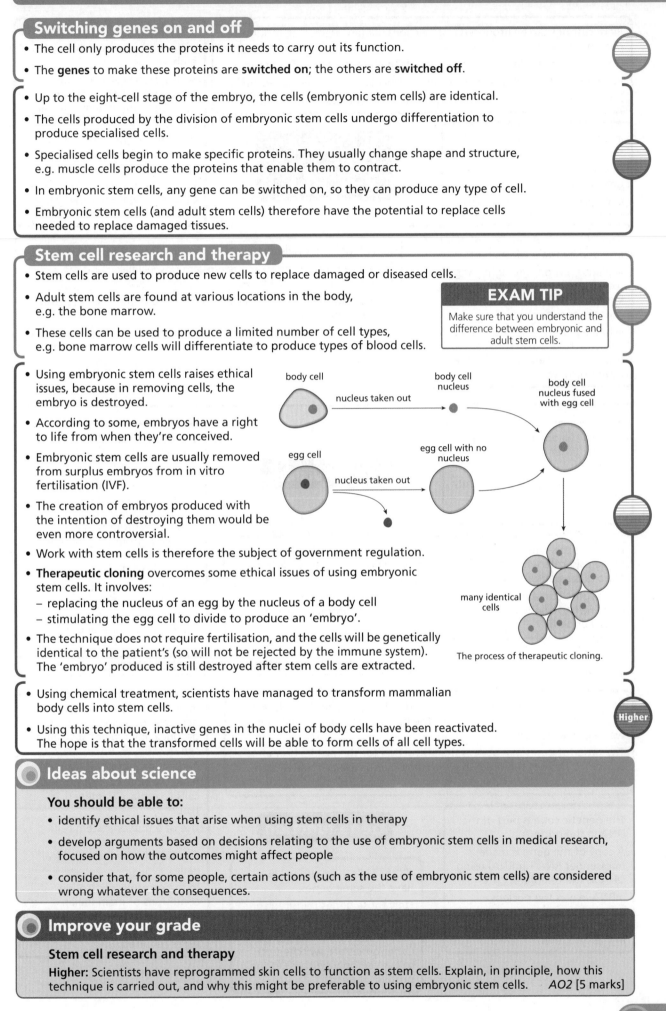

- Using embryonic stem cells raises ethical issues, because in removing cells, the embryo is destroyed.
- According to some, embryos have a right to life from when they're conceived.
- Embryonic stem cells are usually removed from surplus embryos from in vitro fertilisation (IVF).
- The creation of embryos produced with the intention of destroying them would be even more controversial.
- Work with stem cells is therefore the subject of government regulation.
- **Therapeutic cloning** overcomes some ethical issues of using embryonic stem cells. It involves:
  - replacing the nucleus of an egg by the nucleus of a body cell
  - stimulating the egg cell to divide to produce an 'embryo'.
- The technique does not require fertilisation, and the cells will be genetically identical to the patient's (so will not be rejected by the immune system). The 'embryo' produced is still destroyed after stem cells are extracted.

The process of therapeutic cloning.

- Using chemical treatment, scientists have managed to transform mammalian body cells into stem cells.
- Using this technique, inactive genes in the nuclei of body cells have been reactivated. The hope is that the transformed cells will be able to form cells of all cell types.

Ideas about science

You should be able to:

- identify ethical issues that arise when using stem cells in therapy
- develop arguments based on decisions relating to the use of embryonic stem cells in medical research, focused on how the outcomes might affect people
- consider that, for some people, certain actions (such as the use of embryonic stem cells) are considered wrong whatever the consequences.

Improve your grade

Stem cell research and therapy

Higher: Scientists have reprogrammed skin cells to function as stem cells. Explain, in principle, how this technique is carried out, and why this might be preferable to using embryonic stem cells.   *AO2* [5 marks]

Plant cuttings can be produced from sections of plant stems that include parts of meristems.

Plant cuttings can develop into new plants that are clones of the parent.

Rooting of plant cuttings can be promoted by plant hormones called auxins.

Plant cuttings are used to grow plants with identical, desirable features.

A fertilised egg cell (zygote) divides by mitosis to form an embryo.

At the eight-cell stage, all the cells are embryonic stem cells. Embryonic stem cells can produce any type of cell.

After the eight-cell stage, most of the embryo's cells become specialised.

In the adult, some (adult) stem cells remain; these can develop into certain cell types.

In multicellular organisms, cells become specialised.

Groups of specialised cells are grouped into tissues.

Groups of tissues form organs.

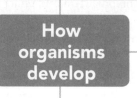

How organisms develop

In plants, only cells in meristems can divide by mitosis.

New cells produced by meristems are unspecialised and can develop into any cell type.

Specialised cells are grouped into tissues, e.g. transporting tissues (xylem and phloem).

Tissues are grouped into organs, e.g. stems, roots, flowers, leaves.

Growth and development of plants is affected by the environment.

A plant's response to light – it grows towards it – is called phototropism. Phototropism increases a plant's chances of survival.

When a plant is exposed to light from one direction, auxin is redistributed towards the shaded side, where it causes growth.

When a cell divides by mitosis, two *daughter* cells are produced.

Each daughter cell is genetically identical to the parent and to each other.

Cell division

The cell cycle involves:
- *Cell growth:* when numbers of organelles (structures in cells having a function) are increased and the DNA molecule of each chromosome is copied, producing 'double chromosomes' with two strands of DNA.
- *Cell division:* the two stands of DNA in each of the chromosomes separate, two nuclei form and the cytoplasm divides.

Meiosis is the type of cell division that produces gametes (sex cells). In meiosis, cells with half the number of chromosomes are produced. Meiosis is essential to produce sex cells.

The zygote contains a set of chromosomes from each parent, so if sex cells were to divide by mitosis, at fertilisation, the chromosome number would double.

DNA is a double helix, like a twisted ladder. The rungs of the ladder are bases. There are four bases in DNA. The bases always pair in the same way – A with T; C with G.

The order of bases in a gene is the instruction for the production of a protein. It defines the order in which amino acids are assembled into a protein.

In embryonic stem cells, any gene can be switched on, to produce any type of specialised cell.

Embryonic and adult stem cells have the potential to produce cells to replace damaged tissues.

Use of embryonic stem cells is subject to government regulation because of ethical issues.

In the carefully controlled conditions of mammalian cloning, it is possible to switch on genes to produce the cell types required.

The genetic code is part of the DNA in the nucleus.

A copy of the gene is made (messenger RNA – mRNA) using DNA as the template.

mRNA leaves the nucleus, attaches to a ribosome, and proteins are assembled on it in the cytoplasm.

Gene function

All the cells in an organism have the same genes.

Only those genes required for a type of cell to function are switched on.

Other genes are switched off, so the cell only synthesises the proteins it needs.

The nervous system

Sending messages

- **Multicellular** organisms need communication systems, so that the body works as a whole and not as individual cells or organs.

- The two communication systems are the **nervous system** and the hormonal system.

- The nervous system:
  - sends messages using nerve cells or **neurons**, which produce a quick, short response; the nerve message, or nerve **impulse**, is *electrical*
  - has specialised organs called the brain and spinal cord.

- The hormonal system produces chemical messages in the form of **hormones**. The system is slower than the nervous system, but the response is longer-lasting.

- In humans and other vertebrates, the **central nervous system (CNS)** consists of the brain and spinal cord.

- In the mammalian nervous system, the CNS is connected to the **peripheral nervous system (PNS)**. This is the neurons, which connect the CNS to the whole body.

- There are different types of neurons:
  - **Sensory neurons** connect **receptors**, e.g. in the eyes, ears and skin, which detect changes in the environment (called **stimuli**), with the central nervous system.
  - **Motor neurons** connect the central nervous system to **effectors**, e.g. muscles, which produce a **response**.

- Hormones are chemicals that are produced by **glands**. They are transported in the blood. This means that all organs of the body are exposed to them, but they affect only their 'target' cells.

- In the hormonal system, responses are slower and longer-lasting. For example:
  - **Insulin** is produced by the **pancreas**. It acts on the liver, muscles and body cells to take up **glucose** from the blood.
  - **Oestrogen** is produced by the **ovaries**. It is a sex hormone that controls the development of the adult female body at puberty, and the menstrual cycle.

central nervous system — brain, spinal cord

peripheral nervous system

The central and peripheral nervous systems.

Remember!
The nervous system uses electrical messages; the hormonal system uses chemical messages.

Neurons

- Neurons are cells **specialised** for carrying nerve impulses, so they are often very long.

- Neurons consist of the **cell body**, which contains the nucleus of the cell, and a long **axon**. Branches on the cell body called **dendrites** receive inputs from other cells (receptors and **nerves**) and conduct impulses towards the cell body.

- Axons carry impulses away from the cell body (to other nerves and muscles).

cell body

nucleus

dendrites

fatty sheath (insulation)

direction of impulse

axon

A motor neuron.

- The axon is a long extension of the **cytoplasm** in a neuron that communicates with the CNS or effector. Some neurons are therefore the longest cells in the body.

- Some axons are covered with an insulating fatty sheath called the **myelin** sheath.

Remember!
Neurons communicate with other neurons, but they do not physically touch each other.

- The speed of the nerve impulse is affected by:
  - temperature (the speed is increased; it's always faster in warm-blooded animals than cold-blooded animals)
  - the diameter of the axon (the wider the axon, the quicker the response)
  - the myelin sheath (as well as insulating the neuron from neighbouring cells, the presence of the myelin sheath speeds up the nerve impulse – it is able to 'jump' from gap to gap along the sheath, making it travel much more quickly).

Improve your grade

Neurons
Higher: Explain how nerve cells (neurons) are adapted to transmitting nerve impulses.   AO2 [5 marks]

Linking nerves together

Synapses

- Some neurons send messages to other neurons. There is a small gap called a **synapse** between one neuron and the next, through which the message has to be transmitted.

- As the nerve impulse reaches the end of the nerve, it is changed to a chemical signal, which crosses the synapse and sets up an electrical impulse in the next neuron.

- Sometimes a neuron has many synapses so that it can communicate information with all these neurons.

- There is no physical connection between neurons. The presence of a synapse means that a nerve is able to communicate better with several neurons that may go to different locations.

- As the nerve impulse reaches the end of the first neuron, a chemical **transmitter substance** is released.

- The transmitter diffuses across the synapse and binds with **receptor molecules** on the membrane on the next neuron. This initiates a nerve impulse in the next neuron.

- After an impulse has been transmitted across, the chemical transmitter is removed from the synapse (it is taken back up by the neuron or broken down by an enzyme).

- There are many different types of transmitter molecules. These work on different nervous pathways, e.g. serotonin is a transmitter that is important in brain function.

- Some transmitters work by inhibiting the next nerve instead of exciting it. Others work on muscles instead of nerves.

- Different transmitters have different receptor molecules.

Higher

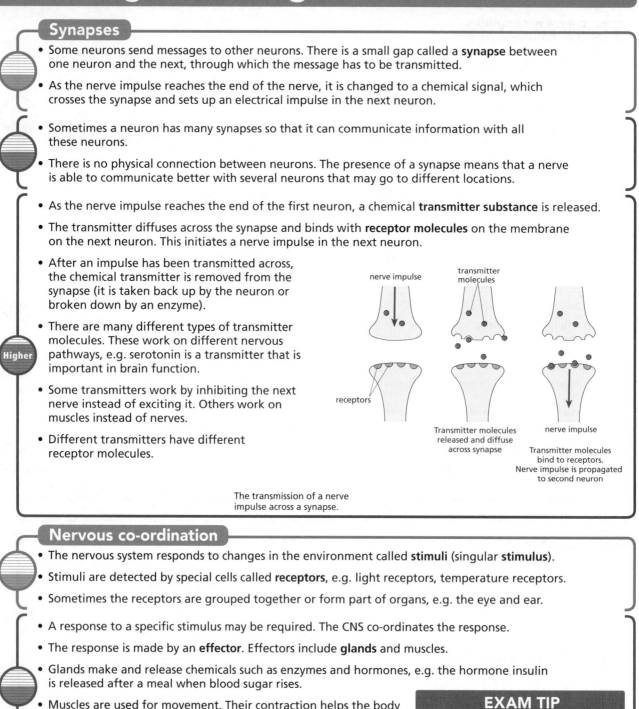

nerve impulse

transmitter molecules

receptors

Transmitter molecules released and diffuse across synapse

Transmitter molecules bind to receptors. Nerve impulse is propagated to second neuron

nerve impulse

The transmission of a nerve impulse across a synapse.

Nervous co-ordination

- The nervous system responds to changes in the environment called **stimuli** (singular **stimulus**).

- Stimuli are detected by special cells called **receptors**, e.g. light receptors, temperature receptors.

- Sometimes the receptors are grouped together or form part of organs, e.g. the eye and ear.

- A response to a specific stimulus may be required. The CNS co-ordinates the response.

- The response is made by an **effector**. Effectors include **glands** and muscles.

- Glands make and release chemicals such as enzymes and hormones, e.g. the hormone insulin is released after a meal when blood sugar rises.

- Muscles are used for movement. Their contraction helps the body to move away from dangerous stimuli and towards pleasant ones. Muscles are also used for movement we're not conscious of, e.g. our heartbeat.

EXAM TIP

You need to be clear about the definitions of receptor cells and effector cells.

Ideas about science

You should be able to:

- identify ethical issues when carrying out investigations on how neurotransmitters work in humans and other mammals

- consider that investigation of these could benefit people with deficiencies in neurotransmitters (acetylcholine in Alzheimer's disease; dopamine in Parkinson's disease), so that research might be justified whatever the consequences.

Improve your grade

Synapses

Higher: Describe how a nerve impulse is transmitted from a sensory nerve to a nerve close to it in a spinal cord.

AO1, AO2 [5 marks]

Reflexes and behaviour

Reflexes

- A **reflex** is a simple response to a stimulus, e.g. removing your foot automatically if you step on a sharp object or a hot one.
- The pathway of a reflex action through the nervous system is called the **reflex arc**.

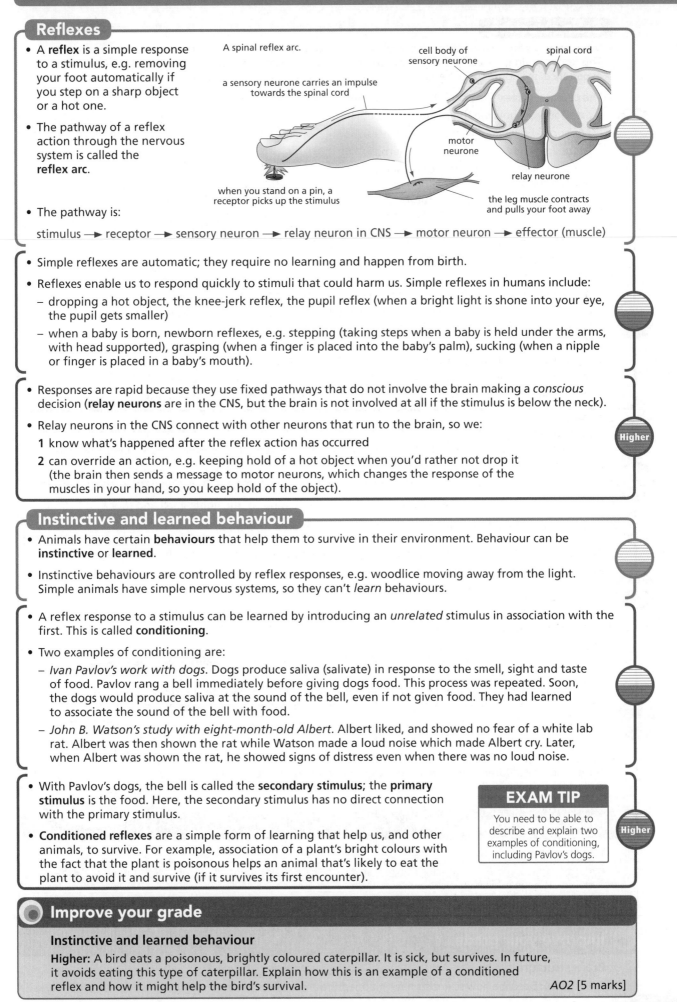

A spinal reflex arc.

a sensory neurone carries an impulse towards the spinal cord

cell body of sensory neurone

spinal cord

motor neurone

relay neurone

when you stand on a pin, a receptor picks up the stimulus

the leg muscle contracts and pulls your foot away

- The pathway is:

stimulus → receptor → sensory neuron → relay neuron in CNS → motor neuron → effector (muscle)

- Simple reflexes are automatic; they require no learning and happen from birth.
- Reflexes enable us to respond quickly to stimuli that could harm us. Simple reflexes in humans include:
  - dropping a hot object, the knee-jerk reflex, the pupil reflex (when a bright light is shone into your eye, the pupil gets smaller)
  - when a baby is born, newborn reflexes, e.g. stepping (taking steps when a baby is held under the arms, with head supported), grasping (when a finger is placed into the baby's palm), sucking (when a nipple or finger is placed in a baby's mouth).

- Responses are rapid because they use fixed pathways that do not involve the brain making a *conscious* decision (**relay neurons** are in the CNS, but the brain is not involved at all if the stimulus is below the neck).
- Relay neurons in the CNS connect with other neurons that run to the brain, so we:
  1 know what's happened after the reflex action has occurred
  2 can override an action, e.g. keeping hold of a hot object when you'd rather not drop it (the brain then sends a message to motor neurons, which changes the response of the muscles in your hand, so you keep hold of the object).

Instinctive and learned behaviour

- Animals have certain **behaviours** that help them to survive in their environment. Behaviour can be **instinctive** or **learned**.
- Instinctive behaviours are controlled by reflex responses, e.g. woodlice moving away from the light. Simple animals have simple nervous systems, so they can't *learn* behaviours.

- A reflex response to a stimulus can be learned by introducing an *unrelated* stimulus in association with the first. This is called **conditioning**.
- Two examples of conditioning are:
  - *Ivan Pavlov's work with dogs*. Dogs produce saliva (salivate) in response to the smell, sight and taste of food. Pavlov rang a bell immediately before giving dogs food. This process was repeated. Soon, the dogs would produce saliva at the sound of the bell, even if not given food. They had learned to associate the sound of the bell with food.
  - *John B. Watson's study with eight-month-old Albert*. Albert liked, and showed no fear of a white lab rat. Albert was then shown the rat while Watson made a loud noise which made Albert cry. Later, when Albert was shown the rat, he showed signs of distress even when there was no loud noise.

- With Pavlov's dogs, the bell is called the **secondary stimulus**; the **primary stimulus** is the food. Here, the secondary stimulus has no direct connection with the primary stimulus.
- **Conditioned reflexes** are a simple form of learning that help us, and other animals, to survive. For example, association of a plant's bright colours with the fact that the plant is poisonous helps an animal that's likely to eat the plant to avoid it and survive (if it survives its first encounter).

EXAM TIP

You need to be able to describe and explain two examples of conditioning, including Pavlov's dogs.

Improve your grade

Instinctive and learned behaviour

Higher: A bird eats a poisonous, brightly coloured caterpillar. It is sick, but survives. In future, it avoids eating this type of caterpillar. Explain how this is an example of a conditioned reflex and how it might help the bird's survival.

AO2 [5 marks]

The brain and learning

Brain structure

- Humans and other mammals have complex brains made up of billions of neurons. This larger brain gave early humans a better chance of survival; it enables learning by experience, including social behaviour, where we are able to interact with others.

- The **cerebral cortex** – the thin, folded, outer layer of the brain – is involved with:
  - *Intelligence* – how we think and solve problems.
  - *Memory* – how we remember experiences.
  - *Language* – how we communicate verbally.
  - *Consciousness* – being aware of ourselves and our surroundings.

  A larger number of folds in the cerebral cortex increases our ability to process information.

The cerebral cortex (the wrinkled surface layer of the brain), which is responsible for conscious thought and actions

The cerebellum, which controls movement and posture

The medulla, which controls breathing and heart rate

The main areas of the human brain.

- **Neuroscientists** map the regions of the brain using invasive and non-invasive methods.

- Invasive methods include:
  - studying how a person is affected when a certain part of the brain is damaged
  - during brain surgery, using electrodes to stimulate parts of the brain electrically, and seeing how patients are affected, including reporting memories and sensations.

- Non-invasive methods include producing images and mapping activity with scanning techniques, e.g. magnetic resonance imaging (MRI). These are useful in:
  - comparing non-diseased brains with the brains of people with brain disease, e.g. Alzheimer's disease
  - looking at activity in the brain when it's stimulated (by music, language or images).

Learning

- Transmitting impulses in the brain leads to links forming between the neurons. This is called a **neuron pathway**.

- If an experience is repeated, more and more impulses follow the same pathway. The pathway is strengthened.

- Neuron pathways are also strengthened by strong stimuli using colour, light, smell and sound.

- Learning happens in the brain as neuron pathways develop in the brain.

- Repeating actions strengthens neuron pathways; we get better at certain skills the more we practise.

- Learning results from experience where:
  - new neuron pathways form (and other pathways may be lost)
  - certain pathways in the brain become more likely to transmit impulses than others.

- Neuron pathways are formed more easily in children than adults.

- With billions of neurons in our brains, the potential number of neuron pathways is huge. This means we can adapt to new situations and respond to new stimuli.

- Children are born with certain instinctive responses to stimuli, e.g. the rooting reflex, where they turn their face towards a stimulus to aid breast feeding, but soon develop learned behaviours.

- **Higher** Children not presented with new, appropriate stimuli, or those isolated during development, may not progress in their learning.

- Evidence suggests that children can only acquire certain skills at a particular age. **Feral** children (children who have lived away from human contact since a very early age) develop only limited language skills when returned to civilisation.

Remember!
It's the interaction between humans and their environment that enables neuron pathways to develop.

Ideas about science

You should be able to:

- identify that some forms of scientific research into the development of learning in humans and other mammals have ethical implications
- consider arguments and actions in ethical issues concerning techniques used to map the human brain.

Improve your grade

Brain structure

Foundation: Describe how scientists have mapped the areas of the brain to see how it works. *AO1* [3 marks]

Memory and drugs

Memory

- **Memory** is the storage and then retrieval (bringing back) of information.
- There are two types of memory:
  - **Short-term memory** involves information from our most recent experiences, which is only stored for a brief period of time.
  - **Long-term memory** involves information from our earliest experiences onwards that can be stored for a long period of time
- You are more likely to remember information if:
  - There is a *pattern* to it. To remember information with no obvious link, you could try to put a pattern to it.
  - You use *repetition* (repeating things), especially over an extended period of time. You could read or rehearse something several times. Evidence suggests that the time intervals between the repeats is important.
  - There is a strong stimulus associated with it. Strong colours, bright light, strong smells or loud sounds associated with information can help us to remember it.

EXAM TIP

You need to understand the terms storage, retrieval, repetition and forgetting when referring to memory models.

- Scientists use models to try to explain how we store and retrieve information.
- The **multi-store model** splits memory into sensory memory, short-term memory and long-term memory, and shows how these work together.
- If information arrives in a memory **store** that is not passed on or retrieved, the information is lost, i.e. forgotten.

The multi-store memory model.

- Models are limited in explaining how memory works. This is because:
  - Memory is more complicated than shown in the model.
  - No models have an exact explanation of how long-term memory works.
  - The multi-store model is too linear, and doesn't provide sub-divisions of short-term and long-term memory.
  - The model does not differentiate between different types of stimulus and the difference in performance of individuals.

Drugs and the nervous system

- Many drugs and **toxins** work by affecting the transmission of nerve impulses across synapses, stopping the transmission, changing the speed of the transmission, or making the impulse stronger or weaker. For example:
  - The **antidepressant** Prozac increases levels of the **transmitter substance** called **serotonin**.
  - Curare, used by South American Indians as an arrow poison, blocks the action of another type of transmitter molecule.

- **Beta blockers** are prescription drugs that block the transmitter molecule **adrenaline**, so they reduce the heart rate. They're used to treat people with problems with their heart rhythm, but some people use them to control anxiety during public performances.

- The drug **Ecstasy (MDMA)** works on serotonin, the same transmitter that Prozac affects.

- Following the transmission of a nerve impulse, the transmitter molecules should be removed from the synapse.
- MDMA blocks the sites on the neuron where MDMA is reabsorbed, increasing its concentration.
- MDMA therefore gives a feeling of well-being, because of increased levels of serotonin.
- After taking MDMA, the brain's serotonin is depleted, so the person is irritable and tired.

Improve your grade

Drugs

Foundation: Some chemicals affect how nerve impulses are transmitted across synapses. Give **two** examples of these chemicals, and state how these chemicals work. *AO1, AO2* [3 marks]

Responding to change

A receptor is used to detect a stimulus; the response is produced by an effector.

Receptors and effectors can form part of complex organs.

The nervous system produces a quick, short-lived response to a stimulus.

The hormonal system produces a slower, longer-lasting response.

The nervous system

The nervous system is the central nervous system (brain and spinal cord) and the peripheral nervous system (the nerves).

The CNS co-ordinates an animal's response to a stimulus.

Sensory neurons carry impulses from receptor cells to the CNS.

Motor neurons carry impulses from the CNS to effectors.

Sensory and motor neurons are linked by relay neurons.

The outer, folded layer of the brain is the cerebral cortex. It is concerned with intelligence, memory, language and consciousness.

Scientists map the brain using several techniques.

The nervous system is made up of nerve cells or neurons; these transmit electrical impulses.

A neuron has a cell membrane and cytoplasm, which is extended into an axon. The myelin sheath (which surrounds some axons) insulates the nerve and speeds up the nerve impulse.

Reflex actions

A reflex arc is a fixed nervous pathway that enables quick, 'automatic' responses independent of the brain.

Simple reflexes include dropping a hot object and newborn reflexes, e.g. sucking.

A spinal reflex arc includes a receptor → sensory neuron → relay neuron → motor neuron → effector.

Synapses

Neurons do not connect physically; impulses are transmitted across gaps called synapses.

Some toxins and drugs affect the transmission of nerve impulses across synapses.

There are many different transmitters, each corresponding to a specific receptor.

An impulse arriving at the end of a nerve causes the release of a chemical transmitter.

The transmitter diffuses across the synapse, binds to receptor molecules, and sets up an impulse in the next neuron.

Learning and behaviour

Simple organisms rely on reflexes for most of their behaviour; these aid survival.

More complex organisms show learned behaviours.

Conditioning is a reflex response to a new (secondary) stimulus, learned by introducing it together with a main (primary) stimulus.

The human brain has billions of neurons that allow learning by experience.

When interacting with the environment, new neuron pathways form in the brain.

New skills can be learnt by repetition (this strengthens neuron pathways).

The number of pathways possible enables mammals to adapt to new situations.

Evidence suggests that children may only acquire some skills at a certain age.

Memory

Memory is the storage and retrieval of information.

Memory can be short-term memory or long-term memory.

The multi-store model shows how short- and long-term memory are linked, but models of memory are limited for a number of reasons.

You are more likely to remember something if you can see, or put, a pattern to it, repeat the information over a period of time, or associate a strong stimulus with it.

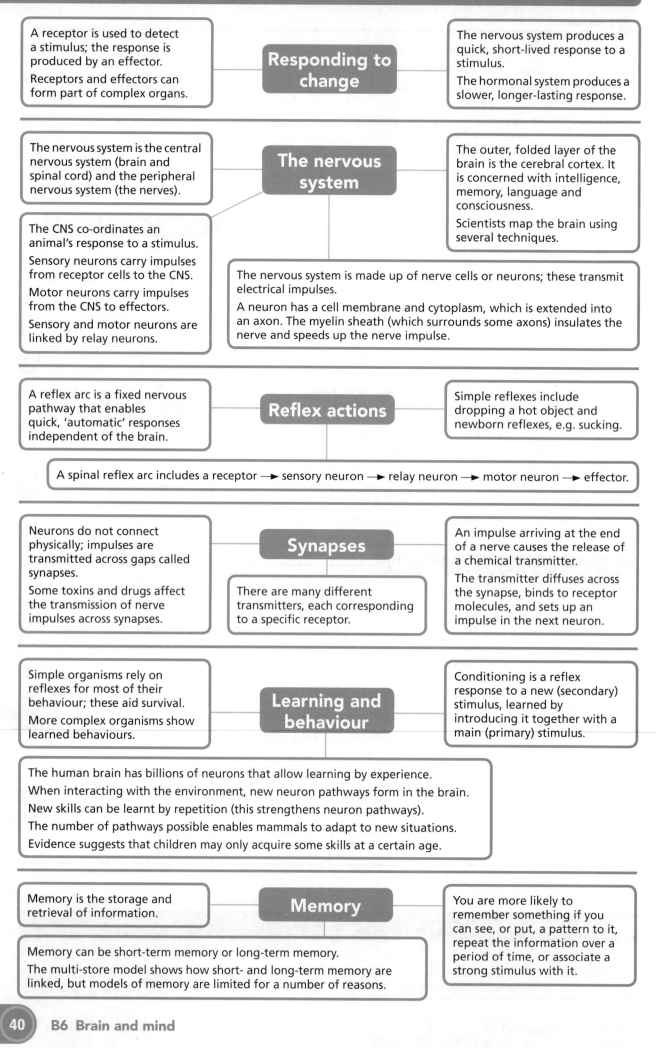

How the body moves

Maintaining posture

- The skeleton has four main functions:
  - Support, the bones are held together with joints and muscles so we remain upright.
  - Protection, bones can protect our internal organs, such as our brain.
  - Blood production, the marrow inside larger bones makes blood.
  - Movement, long bones enable us to move.
- For us to be able to move we rely on muscles to pull on the bones to make them move.
- A muscle contracts, pulling the bone.
- To move a bone back into its original position a second muscle has to pull it back in the opposite direction.

- Muscles **contract** when proteins in their cells react together.
- When contracting, the ends of the muscles are pulled closer together, making the muscle shorter.
- Larger muscles can pull with a greater force than smaller muscles.
- With training, the size of a muscle can be increased and so the force that the muscle can pull will also increase. This is why athletes have larger muscles than non-athletes.
- When a muscle contracts, another muscle relaxes.
- The two muscles involved in the movement of a bone are called an antagonistic pair.
- Most muscles in the body are in an **antagonistic pair**.

Remember!
Muscles can only pull, never push!

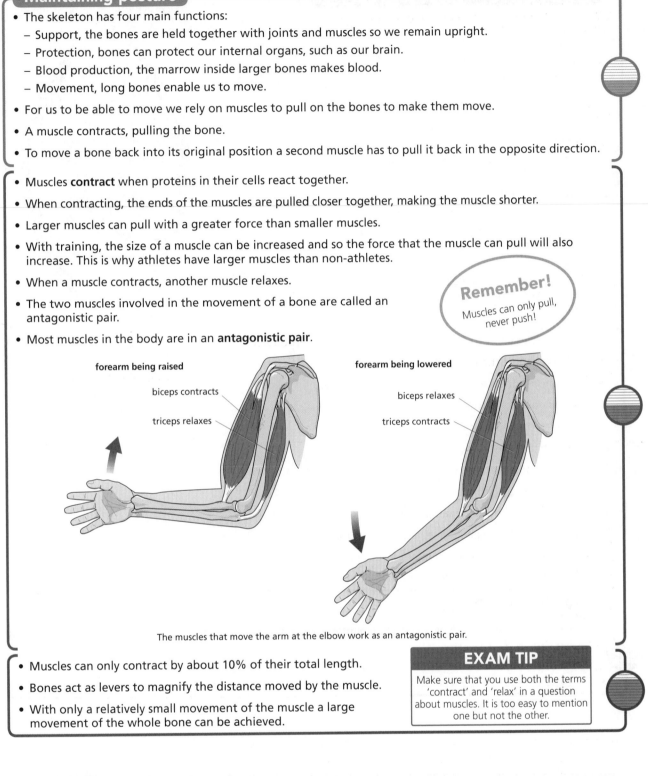

forearm being raised
biceps contracts
triceps relaxes

forearm being lowered
biceps relaxes
triceps contracts

The muscles that move the arm at the elbow work as an antagonistic pair.

- Muscles can only contract by about 10% of their total length.
- Bones act as levers to magnify the distance moved by the muscle.
- With only a relatively small movement of the muscle a large movement of the whole bone can be achieved.

EXAM TIP
Make sure that you use both the terms 'contract' and 'relax' in a question about muscles. It is too easy to mention one but not the other.

Ideas about science

You should be able to:

- Carry out percentage increases in the measured length of muscle before and after contraction.

Improve your grade

How the body moves

Higher: If someone tears their Achilles tendon (the tendon connecting the heel to the calf muscle at the back of the leg), they will be unable to stand on their toes. Explain why this is the case.   *AO1* [4 marks]

Joints

Keeping bones connected together

- Joints are where two bones are connected together.
- Most joints are known as synovial joints.
- **Synovial joints** allow the bones to move relatively freely.
- Shoulders, elbows, hips and knees and the joints in fingers are examples of synovial joints.
- The synovial membrane lines the inside of the joint. It produces the synovial fluid.
- Synovial fluid is slippery and reduces friction in the joint.
- Articular cartilage is shiny, smooth and hard. It stops the ends of the bones being worn away.
- Ligaments can bend and stretch slightly. They join the bones together and keep the joint stable.

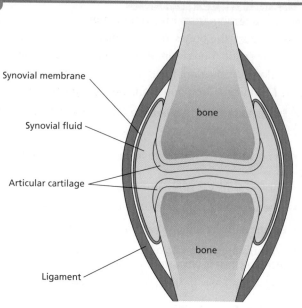

Synovial membrane

Synovial fluid

Articular cartilage

Ligament

bone

bone

The parts found in every synovial joint.

- When muscles contract they pull on the bone with a large force.
- **Tendons** connect the muscle to the bone.
- They transmit the force between muscle and bones.
- Tendons do not stretch.
- Ligaments connect bone to bone, stabilising the joint.
- Ligaments are elastic and so do stretch.
- Without ligaments there would be a danger of the bones coming away from the joint.
- Ligaments can absorb some of the shock of an impact.
- The knee and elbow are examples of hinge joints.
- Hinge joints move in one plane, a bit like a door hinge.
- The shoulder and hip are examples of ball and socket joints.
- Ball and socket joints have a wider range of movement than hinge joints.

> **Remember!**
> It's easy to confuse tendons with ligaments. Tendons are always tight, ligaments are elastic.

Measuring risk

- Problems with a joint can lead to a sports injury.
- Injuries vary in their severity.
- Sports which involve more contact are more likely to result in injury.
- The benefit of carrying out the activity has to be weighed up against the **risk** of injury.

Ideas about science

You should be able to:

- Use the ideas about risk to discuss the chance of a sports injury occurring and the consequences if it did.

Improve your grade

Joints

Foundation: Label a diagram of a knee joint.

AO1 [2 marks]

Exercise and health

BMI as a measure of fitness

- We can use the **body mass index (BMI)** to give an idea of how healthy our weight is.
- BMI is calculated by:

$$BMI = \frac{body\ mass\ (kg)}{[height\ (m)]^2}$$

- The BMI is then checked on a table or graph to see if it is in the healthy range, or not.

BMI value	How is your body weight affecting your health?
17.5 or under	Seriously underweight, significant health risks
17.5 to 19	Underweight, some health risks
19 to 25	Normal, healthy weight
25–30	Overweight, some health risks
More than 30	Obese, significant health risks

- The BMI alone is not enough to determine whether a body weight is healthy or not.
- This is because the proportion of muscle and fat may be different than in a normal body.
- Muscle is denser than fat and so weighs more than an equal volume of fat.
- If you are physically active, you may have more muscle than fat. This would mean that you could have a BMI in the overweight category yet be fully physically fit.

- If your BMI and body fat proportion are too high then you need to lose weight.
- An exercise programme is one way to lose weight.
- Before starting an exercise programme certain factors have to be considered, such as:
  - symptoms
  - current medication
  - alcohol and tobacco consumption
  - current level of physical activity
  - family medical history
  - previous treatments.
- Considering these factors will ensure that the training is as effective as possible.
- If the factors are likely to have an effect, they need to be addressed before the exercise programme is started.
- An exercise programme should start slowly and build up in intensity over time.
- This is to prevent excessive strain on the heart, muscles and joints.

- We can take measurements of our weight to check progress.
- The accuracy of the data obtained depends upon the accuracy of the monitoring technique (e.g. bathroom scales) and how often the data is repeated.
- There will always be natural variation in results.
- A range of data is needed to show when there has been true progress.
- Data is best presented as **averages**.

Higher

Ideas about science

You should be able to:

- Explain why repeating measurements leads to a better estimate of the quantity.
- Suggest reasons why several measurements of the same variable may give different values.

Improve your grade

Exercise and Health
Foundation: Why is BMI used to decide whether someone is overweight or not?   *AO1 [3 marks]*

Exercising safely

Heart rate and exercise

- Exercise leads to an increase in heart rate (how often the heart beats per minute) and blood pressure (the pressure of the circulating blood on the blood vessels).

- An increased heart rate supplies more blood to the muscles.

- Blood delivers oxygen and glucose for respiration and removes waste products such as carbon dioxide.

- Respiration converts glucose and oxygen to carbon dioxide, and gives out energy which the cells use.

- Muscle cells use the energy to contract.

- If the heart rate is too high the muscles can become strained and blood pressure can increase.

- If blood pressure is too high then delicate blood vessels in the eye or heart can burst.

- For safe exercising the heart rate needs to be between 70% and 90% of the maximum rate your heart can actually achieve.

- The formula for your approximate maximum heart rate (in beats per minute) is:
220 – your age in years.

- The **recovery period** is the time taken after exercise for the heart to return to a normal rate. The fitter you are, the quicker the recovery rate.

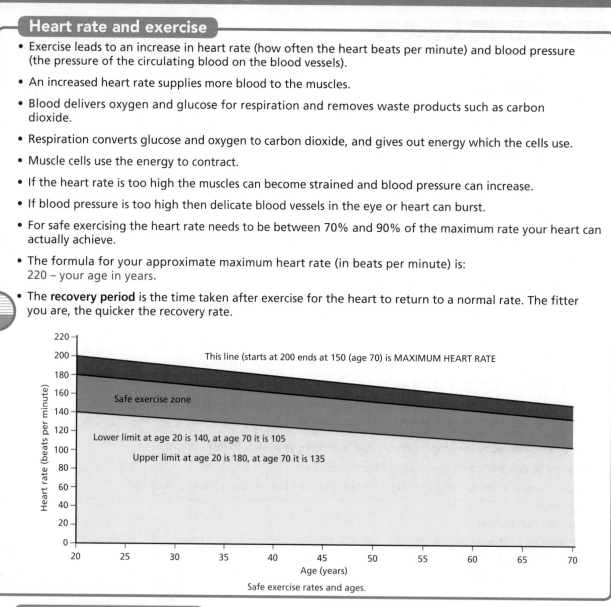

Safe exercise rates and ages.

Injuries and treatment

- Injuries take place when the body is exercised more than it can physically take.

- **Sprains** occur when ligaments overstretch and so the joint becomes more wobbly.

- Torn ligaments and tendons are where the ligament or tendon becomes disconnected.

- A **dislocation** is where a bone, such as the hip bone, comes out of its socket.

- The treatment for a sprain is PRICE: Protect, Rest, Ice, Compression and Elevation.

- It is important to make sure that the patient is suffering from a sprain and not a more serious injury before treating the injury.

- A **physiotherapist's** role is to help an injured person strengthen any damaged joints by providing a suitable exercise programme.

Improve your grade

Exercising safely

Foundation: Three competitors take part in a marathon. Arwen, Roz and Peter are aged 32, 45 and 65 respectively. Show by calculation their respective maximum heart rates and give the highest heart rate they should exercise at.

AO2 [3 marks]

How the heart works

Circulating blood

- The human heart has four chambers joined together within one large block of muscle.

- The heart has four tasks:
  - To collect deoxygenated blood at low pressure from the body.
  - To pump deoxygenated blood to the lungs.
  - To collect oxygenated blood at low pressure returning from the lungs.
  - To pump oxygenated blood at high pressure to the rest of the body.

- Each of these tasks is undertaken by a pumping action of the heart.

The human heart.

- Humans have a **double circulation system**.

- Blood pumps the blood twice for each trip around the whole body.

- Deoxygenated blood enters the heart on the right-hand side via the vena cava.

- It then moves from the right atrium into the right ventricle and then into the lungs via the pulmonary artery.

- The blood slows down when moving through the lungs to enable gas transfer.

- Carbon dioxide leaves the blood and oxygen enters the blood.

- Gas transfer is via diffusion from high to low concentration.

- The blood, now oxygenated, moves back into the heart via the pulmonary vein.

- It then moves into the left atrium and then the left ventricle before being pumped at high pressure to the rest of the body from the aorta.

- The heart forms part of the circulatory system. This is a collection of blood vessels, including arteries, veins and capillaries.

> **Remember!**
> When you look at the heart drawn on the page, the right-hand side of the heart is on the left-hand side of the exam paper.

	Arteries	Capillaries	Veins
Walls	Thick, with layers of muscle and elastic fibres	Thin, single celled	Thin, with no muscles
Direction of blood flow	Away from the heart	Through tissues	Towards heart
Blood	Oxygenated	Oxygenated at start, deoxygenated at end	Deoxygenated
Valves	Not present	Not present	Present

EXAM TIP

Make sure that your definition for double circulation clearly indicates that the heart pumps twice for every single trip of blood around the body.

Valves

- **Valves** in the heart and veins prevent the backflow of blood.

- In the heart the valves close with an audible noise. This can be detected using a microphone.

- If there is a faulty valve this can cause problems, as the blood will not reach the lungs to be re-oxygenated.

- In the veins valves prevent gravity from keeping the blood in the lower part of the body.

- Varicose veins are where the valves stop working and cause the blood to pool at the next working valve down.

- The aorta and pulmonary arteries are the only arteries in the body that have valves.

Improve your grade

How the heart works

Higher: Suggest why patients in a persistent vegetative state have to be turned regularly in their beds.

AO2 [4 marks]

Blood components

What's in blood?

- Blood is made up of four different components.
- Red blood cells:
  - are full of **haemoglobin**, which binds with oxygen
  - have no nucleus, to give more space for haemoglobin
  - carry oxygen around the body for respiration by cells
  - have a smooth shape to allow them to slip along narrow capillaries.
- White blood cells:
  - come in different versions
  - some recognise invading organisms, such as viruses and bacteria, and swallow them up
  - some produce antibodies that can quickly identify invading organisms in the future.
- Platelets:
  - help the blood clot at the sites of injury.
- Plasma:
  - is a fluid that carries nutrients, antibodies, hormones and waste substances around the body.

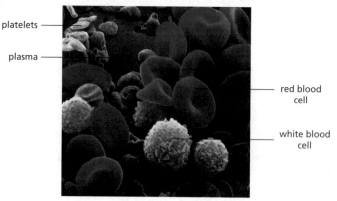

The components of blood

- 90% of plasma is made of water.
- The remaining 10% of plasma is made up of:
  - **glucose**, to provide energy for all body cells through respiration
  - proteins, such as antibodies, to help protect the body
  - salts to keep the blood at the right concentration
  - hormones to help control a range of functions in the body, such as growth, puberty and blood sugar levels
  - waste substances, such as carbon dioxide, which is transported to the lungs, and urea, which is carried to the kidneys.
- The platelets in the blood start a series of reactions at the site of injury.
- This produces a mass of protein called fibrin, which stops the bleeding.

Tissue fluid

- When the blood reaches the capillaries, the liquid plasma, rich in oxygen and dissolved food, leaks out into the space between them.
- The spaces between the capillaries and the cells is called the capillary bed.
- The dissolved glucose and oxygen can then move into cells by diffusion.
- Waste products, such as urea and carbon dioxide, can diffuse back into the blood in the capillary.
- White blood cells can also escape into the tissue.
- Most of the fluid passes into tubes called lymph vessels.

Higher

Improve your grade

Blood components

Higher: Lorna is in training for the Rio Olympics. She trains at a high altitude, which increases the red blood cell content of her blood. Explain how this may give her an advantage over other competitors.

AO2 [6 marks]

Blood as a transport system

Carrying oxygen

- Red blood cells are packed full of an iron-containing chemical called **haemoglobin**.

- Oxygen moves from a high concentration in the lungs to a lower concentration in the red blood cells.

- This forms **oxyhaemoglobin**.

- When the red blood cells move past body cells that have a lower level of oxygen, such as active muscle cells, the oxygen leaves the oxyhaemoglobin and diffuses into the cell.

- Plasma in the blood absorbs the carbon dioxide as well as urea waste from the cells.

- The carbon dioxide leaves the blood plasma in the lungs.

Gas exchange in the lungs.

- Blood does not just transport dissolved gases.

- Some white blood cells travel in the blood to protect against infections.

- Other white blood cells can engulf and consume invading microorganisms.

- Another class of white blood cells can release antibodies that mark out foreign bodies, targeting them for other white blood cells to destroy.

- The plasma transports essential chemicals around the body.

- Glucose is vital for respiration, giving cells energy.

- Waste products travel in the blood plasma.

- Urea, made by the liver from waste amino acids, is transported to the kidneys.

- Kidneys filter the blood, moving the urea into urine, which can be excreted.

- Hormones are chemicals that influence the functions of different cells in the body.

- Insulin is a hormone made in the pancreas and is transported by the plasma.

- Insulin influences uptake of sugar by cells.

- Adrenaline is a hormone that prepares the body for vigorous activity.

- It is produced by the adrenal glands near the kidneys and transported in the plasma.

Remember!

Only oxygen is carried by red blood cells. It's the plasma in the blood that carries all the other chemicals.

EXAM TIP

Never write a white blood cell 'fights' a virus. Examiners do not like the word, as a fight can be lost. It is better instead to say 'kill' or 'destroy'.

What makes red blood cells so effective?

- Red blood cells have no nucleus, unlike all other cells.

- This means that the whole cell can be packed with haemoglobin.

- Red blood cells also are a biconcave shape.

- This maximises the surface area of the cell, which means that the highest rates of gas exchange can be reached.

- The circular shape fits perfectly in tubes. With no sharp edges, blockages of the blood vessels are very unlikely.

- The disorder sickle cell anemia leads to curved, sickle-shaped red blood cells.

- Sickle-shaped cells are far more likely to clump together and block the blood vessels.

Higher

Improve your grade

Blood as a transport system

Foundation: Mark wants to have his DNA tested. Which component of his blood will be used to test his DNA? Explain your answer. *AO1* [3 marks]

Keeping cool

Controlling body temperature

- Humans need to keep their body temperature at about 36.9°C.
- If the temperature gets much higher, then chemical reactions in the body start to go faster.
- This can cause serious problems, even death.
- If you are working hard in high temperatures, then the problem will be worse.
- In order to reduce body temperature:
  - blood is diverted to the skin surface; this allows heat to be radiated away from the skin
  - sweat is produced; this cools the skin down as it evaporates.
- Sweating cools the body very effectively, but loses fluid.
- The fluid that is lost is made up of water, urea and salts.
- If the salts and water are not replaced, then the body becomes **dehydrated**.
- Sweating will eventually be reduced to save body water and can even stop altogether.
- This means the body will get hotter.
- This puts even more stress on the organs of the body – eventually leading to death.

45°C	Death
40°C	Unconsciousness
39°C	Person becomes delirious, often vomits
36.9°C	Normal body temperature
35°C	Person feels cold, shivering begins
33°C	Uncontrollable shivering, clumsy movements and pain
30°C	Shivering slows or stops, speech becomes slow, person becomes confused and finds it difficult to stay awake
25°C	Breathing and pulse very slow, heart attacks can occur
20°C	Death

Body responses at different temperatures.

- Sensors in the skin detect external temperatures.
- The signals from the sensors are sent to the brain very quickly.
- Another sensor is located deep inside the brain.
- This directly measures the temperature of the blood flowing through the brain, in other words the internal body temperature.
- The brain acts as a control centre, processing the sensory temperature information from the different sensors and coordinating the response.
- When the body temperature rises the brain sends nerve impulses to the sweat glands, telling them to produce sweat.
- As the water evaporates it cools the skin.
- The water in the sweat uses the heat energy in the skin to change from liquid water to water vapour.
- As blood flows through the skin, more excess heat is released.

Vasodilation and vasoconstriction

- To increase heat loss.
  - As body temperature rises, blood is shunted to the outer layers of the skin.
  - Small blood vessels that supply the capillaries of the skin become wider.
  - This allows more blood to flow near the surface of the skin.
  - The skin appears redder and feels warmer as a result.
  - This process is called **vasodilation**.
- To reduce heat loss.
  - The small blood vessels get narrower.
  - This restricts the blood flowing close to the surface of the skin.
  - The skin appears bluer and feels cooler.
  - This process is called **vasoconstriction**.

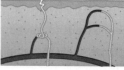

sweat evaporates from the skin surface, cooling it

Vasodilation
When the body is too hot: blood capillaries in the skin dilate and so blood flow increases, bringing more blood to the surface, where it loses heat.

Vasoconstriction
When the body is too cold: blood capillaries in the skin constrict and so less blood flows through them, reducing heat loss.

Vasodilation and the opposite process, vasoconstriction.

Improve your grade

Keeping cool

Foundation: Niamh is suffering from heat exhaustion. Her friend Amy thinks that she is probably alright because she is not even sweating. Give reasons why Amy is making an incorrect diagnosis. *AO2* [4 marks]

Keeping warm

Core and shell temperature

- The core body temperature needs to stay at about 36.9°C.
- The **shell** of the body, including the arms and legs, can get much cooler.
- The skin can drop to below 25°C with little, or no, damage.
- Sensors in the skin pass nerve impulses to the brain to warn of lower temperatures.
- The brain acts as a control centre and responds by:
  - causing the muscles to start shivering
  - diverting blood away from the cold shell into the warmer body core
  - stopping any sweating.

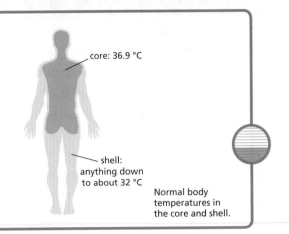

core: 36.9 °C

shell: anything down to about 32 °C

Normal body temperatures in the core and shell.

Controlling the body's responses

- The brain acts as the control centre to coordinate the body's response to low temperature.
- The part of the brain involved in temperature regulation is located near the centre of the brain, towards the base of the brain.
- The brain analyses the inputs from:
  - temperature sensors in the skin
  - measurements of the blood temperature in the brain.
- The brain then coordinates the response, sending impulses along nerves.
- These travel to the **effectors** that will carry out the response, including:
  - muscles (for shivering and raising hairs)
  - sweat glands (to switch off sweating).
- This all takes place unconsciously (without thought).

To summarise:

- If the body temperature is cooler than normal:
  - sweating is stopped
  - skin blood vessels are narrowed
  - blood is kept closer to the core
  - warm clothes may be put on
  - the person may move to a warmer location
  - heat production is increased through:
    - shivering
    - respiration of extra sugar.

- If the body temperature is warmer than normal:
  - sweating is initiated
  - skin blood vessels dilate
  - blood flows closer to the skin
  - clothing can be removed
  - the body can be rested
  - move to a cooler area.

- When the core body temperature falls to below 35°C the person will suffer from **hypothermia**.
- The blood has moved away from the shell to the core.
- People suffering from hypothermia need to be warmed up, but not too quickly.
- If the body is warmed too quickly, the blood can leave the core, to be replaced by much cooler blood. This can cause heart and organ failure and then death.
- The part of the brain responsible for monitoring blood temperature is called the **hypothalamus**.
- Effectors such as muscles and sweat glands work **antagonistically** to alter temperature.
- This means they have opposite effects on body temperature.

Improve your grade

Keeping warm

Higher: Explain the role of the hypothalamus in temperature regulation.

AO1 [4 marks]

Diabetes

Sugar levels in the blood

- When carbohydrates are broken down in the gut they produce sugar.
- The sugar is called **glucose**.
- The sugar passes quickly into the body.
- This causes the blood sugar level to rise.
- A rise in blood sugar level can lead to:
  - increased blood pressure
  - glucose to leak into the urine.
- The increase in blood pressure can cause problems with the kidneys, the eyes and the circulation in the body.
- Losing glucose in the urine is also wasteful.
- Highly processed foods often contain high levels of sugar, which is then rapidly absorbed into the blood.
- The body reacts to a rise in blood sugar level by producing a hormone called **insulin**, in the **pancreas**.
- Insulin works by letting cells in the body take glucose out of the blood into the cell.
- Inside the cell the glucose is used in respiration or converted into another substance for storage.
- When insulin levels drop below a particular level, glucose is no longer absorbed into the cells.
- If someone has diabetes, then they will not be able to control the amount of sugar in the their cells.
- In a person with diabetes the amount of sugar in the blood can vary widely.
- When the level falls too low, the person can go into a very deep sleep called a **coma**.
- There are two types of diabetes:

	Type 1	Type 2
How common is it?	10% of all cases	90% of all cases
When does it start?	Usually in childhood	Often after 40 years of age
How is the body affected?	Body cannot produce insulin	Body cannot produce enough insulin or the body cells stop responding to the insulin produced
What causes the disease?	Not sure, there may be a genetic component	It seems to be linked to a high-sugar diet or being overweight
Treatment	Injections of insulin	Improving the diet by eating foods high in fibre and complex carbohydrates (found in fruit and vegetables), and taking more exercise

- The risk factors for Type 2 diabetes include:
  - being overweight
  - taking little or no exercise
  - being over 40 years of age
  - high blood pressure
  - family history of diabetes
  - unhealthy diet, too much fat and refined sugar, not enough fibre.

Remember!
You will need to be able to interpret data on the risks associated with an unhealthy lifestyle.

Feedback mechanisms

- The control of body temperature and blood sugar levels are examples of **negative feedback**.
- This is where the body responds to a change by doing something that reduces the effect of that change.
- Negative feedback systems allow for a fine-tuned response.

Improve your grade

Diabetes

Higher: Explain why someone who has Type 2 diabetes does not always / necessarily need to inject insulin, whereas a person with Type 1 diabetes does.
AO1 [4 marks]

Cycles in nature

Open and closed loops

- Products are made from different resources.
- There are two types of process to describe how environmentally friendly the process is.
- An open loop system is one where once waste leaves the loop it cannot be used again, e.g.
  - Plastic is made from oil.
  - When a product made from oil is thrown in the bin it could be either:
    - buried in landfill
    - incinerated.
- If the plastic is buried or incinerated it cannot be reused.
- The process of making the plastic, from the oil to its eventual disposal, is therefore an **open loop process**.
- A closed loop system is one where waste can be used again to keep the loop going round for ever, e.g.
  - Books are made from paper.
  - Paper was originally wood from a tree.
  - The wood was pulped and then turned into paper.
- Once paper has been finished with it can be collected and recycled.
- Paper can be processed to make new paper or composted and used to provide soil for more trees to grow.
- This is an example of a **closed loop process**.

> ### EXAM TIP
> Don't confuse the words recycle and reuse. Recycle means to break down and use as a raw material again. Reuse means to clean and then use again for the same or a similar purpose.

- There are many examples of closed loops in the natural world.
- Carbon dioxide is taken in by plants to be used for photosynthesis, using sunlight, a sustainable source of energy.
- The sugar formed is used for energy and to make other compounds containing carbon.
- The material made is called **organic matter**.
- The waste product from photosynthesis is oxygen.
- Oxygen is used by organisms from respiration.
- Animals eat the plants and use the organic material for energy.
- The plants are an **input** for the animals.
- Animals produce two **outputs**.
  - Carbon dioxide from respiration – which can be used as an input by plants.
  - **Faeces** as waste – the chemicals provide an input for microorganisms in soil.
- When animals and plants die their remains provide a source of organic matter, an input, for microorganisms in the soil.
- The microorganisms break down the organic matter for energy and give out carbon dioxide and mineral nutrients.
- Microorganisms also secrete digestive enzymes that can break down complex materials, such as cellulose and lignin, into simpler materials.
- These are then used by plants to grow.
- The cycles are all closed and interlink.

> **Remember!**
> In your exam, you will need to interpret diagrams on a closed loop system through an ecosystem

- Millions of years ago, plants harnessed the energy from the sun through photosynthesis.
- Crude oil was originally formed from plants and animals buried millions of years ago.
- Humans are now burning crude oil, but it cannot be replaced.
- The time scale for the loop to be closed is over a million years, but the damage to the atmosphere caused by the carbon dioxide being released will happen much more quickly.
- On a human time scale it is behaving like an open loop system.

Improve your grade

Cycles in nature
Foundation: Explain why the manufacture of paper is a closed loop process.

AO1 [2 marks]

Cycles in rainforests

Overproduction

- A fruit is a structure that develops from a fertilised flower.

- It contains seeds that grow into a new plant.

- Seeds usually have their own energy store to help the seeds to germinate.

- We eat fruits because they tend to hold more energy than the other parts of the plant.

Seed

Leaf

water · protein · carbohydrate · fat

Nutrient composition of plant parts.

- **Reproductive structures** include:
  - eggs
  - sperm
  - pollen
  - flowers
  - fruit.

- Overproduction of reproductive structures is necessary to ensure that enough survive to grow into a sexually mature adult of that species.

- In trees, such as the Brazil nut tree in Amazonia, this means producing enough structures each year to compensate for the number that will be lost due to:
  - failure of seed germination
  - damage by fungal or insect attacks
  - animal feeding
  - death of the tree before its reproductive age.

- In animals overproduction of sperm and eggs also occurs.

- Loss could be due to:
  - failure of eggs and sperm
  - animal feeding
  - loss of the animals before they reach reproductive age.

- The production of an excess amount of reproductive structures has to be balanced with the energetic cost of the structures.

- Any reproductive structures that are not used are recycled in the ecosystem, so from an ecosystem perspective, the excess production is not wasteful.

Production in rainforests

- The materials in a rainforest cycle through a number of closed loop systems.

- These loops together produce a **stable system**.

- This means the inputs to the rainforest are broadly balanced by the outputs.

- There will always be some losses at the edges of the ecosystem, e.g.
  - animal migration
  - rivers moving minerals from uphill areas.

- The systems are not permanent. They can be disrupted, e.g. by
  - climate change
  - human intervention through:
    - logging
    - removing fruits and other resources
    - manufacturing products
    - building houses.

- There is an ethical dilemma over the needs of local communities and the need to conserve the rainforest.

Improve your grade

Cycles in forests

Foundation: Frogs produce a large number of frogspawn. Suggest why they do this, rather than produce one or two offspring.
AO2 [4 marks]

Protecting soil

Desertification

- Most plants need to grow on soil.
- Soil is made up of a mixture of:
  - minerals
  - organic material, alive and dead
  - air spaces
  - water.
- A good soil provides the right minerals that roots can take up and offers good drainage, so that plants are not in water.
- The plants growing in soil affect it in three ways:
  - the roots hold the soil together
  - the leaves and stem slow down rain so that it hits the ground with less force
  - water is removed from the soil by the plant roots.
- A healthy soil also helps to prevent floods.
- The soil absorbs water slowly, releasing it gradually so that rivers and streams drain away safely.
- If the vegetation is removed then the bare soil is far more likely to become eroded.
- Erosion is through wind and rain. The topsoil gets blown or washed away.
- It is then impossible to plant new crops.
- The area affected can become a desert.

- Rainforests are very complex **ecosystems**.
- They help protect and enrich the soil by covering it with many layers of foliage and having extensive root systems to bind it together.
- The effect of rainforest vegetation is so successful that the water flowing out into rivers at the base of hills will be clear.
- This indicates that the minerals are being retained in the soil.
- Logging of a rainforest removes this protection and a lot of soil is washed into the river.
- Downstream the rivers become clogged and then flood, causing more damage as soil on the banks is washed away.
- Water taken into the plant eventually leaves the plant via a process called **transpiration**.
- Water evaporates from the leaves and re-enters the atmosphere.
- With an ecosystem as large as a rainforest enough water evaporates to make rain clouds.
- If the rainforests are removed weather patterns become more unpredictable.

- Ecological services are functions of the ecosystem that the ecosystem provides for all members of the system.
- There are four areas of service:
  - supporting services, such as: nutrient cycling, oxygen production, pollination, soil formation and protection
  - provisioning services, such as: food, fibre, fuel and water
  - regulating services, such as: climate regulation, water purification and flood prevention
  - cultural services, such as: education and recreation.
- Ecosystem services are provided by different elements of the ecosystem working together.
- Humans take advantage of the ecosystems and the services they provide, e.g.
  - removing water for our own use
  - using wood for building and fuel
  - using resources for food.
- We must be careful not to damage the ecosystem so much that the services become unavailable.

Improve your grade

Protecting soil
Foundation: What are the four main components of soil?   AO1 [3 marks]

Ecological services

- Organisms depend on **ecological services** to survive.

- Ecological services are where the needs of the organism are provided by the actions of other organisms.

- It is a bit like humans needing water, electricity and telephone services connected in the home.

- Examples of ecological services include:
  - A plant depending on an insect to pollinate its flowers. Without the insect, the plant would not be able to reproduce.
  - A beaver needs certain types of trees to provide materials to build its dam.
  - Plants rely on microorganisms breaking down dead organic matter so that it can be absorbed by the plant roots.
  - Honey bees need flowers to produce nectar so that they can make honey to feed the colony.

- The environment in the tropics is warm, wet and sunny.

- This helps plants to grow very fast.

- It also means that microorganisms in the soil can break down dead plants and animals very quickly, releasing their stored minerals.

- This means the minerals enter the soil very quickly.

- As the conditions for plant growth are so good, they can absorb the minerals in the soil rapidly.

- This means that although the cycling is good, the soil is lacking in nutrients (as the plants have absorbed them).

- By comparison a **temperate** climate has less sunshine, is cooler and drier.

- This means that the plants grow more slowly and microorganisms break down dead plants and animals more slowly.

- The nutrients therefore stay in the soil for longer.

- The soil is more fertile, but the cycling is not as good.

- With both the tropics and temperate regions, inputs and outputs are balanced.

- The **slash-and-burn** method of agriculture can alter a rainforest system.

- When the trees are removed (e.g. to make farmland, remove wood for manufacture etc.) this changes the balance of the ecosystem.

- In the first year after removing trees there will be an excess of minerals in the soil that can be exploited.

- After the minerals have been removed, by crops or livestock, the land becomes barren.

- A complex rainforest is replaced by a simpler grassland ecosystem.

- The ecological services provided by the rainforest are lost.

- Desertification can occur as the protective vegetative layer is lost from the soil.

- **Biodiversity** is a measure of the variety of organisms living in a given area.

- Rainforests are the most biodiverse ecosystems on the planet.

- Once the rainforest has gone it is lost forever.

- **Sustainable** forestry means managing the removal of key services.

- This means planning the steady replacement of resources so that the system is only minimally affected.

Improve your grade

Ecological services

Higher: Explain why soil in the UK may be better quality than the soil in the Amazon rainforest.

AO1 [4 marks]

Poisoned lakes

Movement of waste

- Waste that is not recycled has to go somewhere.
  - Domestic refuse is often buried in landfill.
  - Faeces are cleaned from waste water at sewage works and the remains passed into rivers and the sea.
  - Waste gases are passed into the atmosphere and spread out to safe levels.
- Eventually the rubbish is either broken down or spread out so much that it becomes safe.
- The damage done by waste depends on:
  - how much has been produced
  - how dangerous it is
  - how quickly the environment can deal with it.
- Sometimes the environment cannot cope with the waste quickly enough.
- This means that the waste will build up to harmful levels.
- An example is human faeces in sewage. The faeces will break down in the environment but it does take time.
- **Sewage** is classed as anything that gets flushed down the toilet.
- In London, sewage used to drain directly into the River Thames.
- This was not a problem when the population of London was small.
- As the population increased, the sewage built up in the river.
- This then killed the river life, such as fish.
- Some sewage was even able to get into drinking water supplies.
- When drinking water is affected, this increases the spread of diseases such as typhoid and cholera.

- Human sewage is a mixture of complex chemicals.
- It is particularly high in nitrogen-containing compounds.
- The nitrogen is used by microorganisms for growth.
- The number of microorganisms increase and so they use more oxygen.
- The available oxygen decreases as they respire.
- Larger organisms in the river, such as fish, start to die and decay, further lowering oxygen levels.
- The amount of nitrogen compounds continues to increase.
- The river becomes suitable only for anaerobic organisms.

Eutrophication and bioaccumulation

- When compounds that can cause the increased growth of microorganisms enter lakes and rivers it can cause a process called **eutrophication**.
- Chemicals causing eutrophication can come from sewage discharge or from excess organic and inorganic fertilisers applied to crops flowing into rivers.
- The bacteria and algae in the water show a sudden growth, causing **algal blooms**.
- The decay of dead algae from the algal blooms as well as other plant material can quickly use up all the available oxygen in a river or lake.
- Some species of algae can produce toxins, which kill organisms, such as cows, drinking from the river or lake.
- **Bioaccumulation** is where a chemical that enters the environment enters the food chain.
- The amount of the chemical is low at the early stages of the chain.
- However, it is not broken down, so remains inside the body.
- As the next trophic level consumes members of the previous level, they take in more of the chemical, increasing the concentration.
- The highest concentration of chemical will be in the organism at the top of the food chain.
- Depending upon the chemicals' toxicity, it will have potentially harmful effects.

Higher

Improve your grade

Poisoned lakes

Higher: Explain the term eutrophication.

AO1 [3 marks]

Sustainable fishing

Farming fish for food

- Fish are an excellent source of protein.
- Fish can be grown in ponds.
- Carefully measured amounts of the run-off from fields are added to the water.
- The run-off contains fertiliser that was not used by the crop plants.
- This is then used by water weed for growth.
- Fish eat the weed, then grow.
- The ponds are restocked with young fish every year.
- The advantage of this system is the fishermen know:
  - how many fish were put into the pond
  - how many fish they can take out.
- **Biomass** is the total dry weight of the fish in the pond.
- The ponds have a very high biomass compared to lakes and rivers.

- Lakes in Scotland are called lochs.
- Salmon farmers keep fish in giant tanks in lochs.
- The tanks are made of a fine net-like mesh. This keeps the fish in and allows the loch water to flow through.
- To stock the farm, eggs and sperm are removed from adult fish and mixed in the laboratory.
- The fertilised eggs are hatched in tanks of gently flowing water.
- The young fish (called fry) are added to the giant tanks in the loch.
- Food suitable for salmon growth is added.
- When the salmon reach a suitable size, they are harvested.
- The **yield** is the amount of fish produced each year.
- The giant tanks are densely stocked with fish, so the yield is very high.
- Keeping such high numbers of fish close together means **pesticides** need to be added to prevent infections.
- If not managed correctly, the pesticides can leak out of the tanks and damage other wildlife in the loch.
- The Chinese often farm fish in man-made ponds, which are like large fish tanks.
- Fish produce waste. In the salmon farms the waste is washed away naturally. With the Chinese ponds this does not occur as they are an enclosed system.
- Chinese ponds have to be drained, cleaned and restocked every six years.

Fishing in the open sea

- Stock biomass is the amount of fish old enough to produce eggs, which then become the next generation of fish.
- Declines in fish numbers are often due to **overfishing** or rising sea temperatures.
- Overfishing means more fish are removed than are replenished.
- To prevent overfishing, governments agree on fishing quotas.
- A fishing quota controls:
  - the number of days a fishing fleet can fish each year
  - the minimum size of fish that can be taken (smaller ones must be thrown back into the sea).

Improve your grade

Sustainable fishing

Higher: A new salmon farm is proposed for a lake in the Lake District. Suggest a biological reason why some people may protest against it.   *AO2* [4 marks]

Producing protein and penicillin with microorganisms

Farming with microorganisms

- Microorganisms can be used to make many useful products.

- Microorganisms often have a number of useful features for farming.
  - They have a relatively simple biochemistry.
  - They reproduce very quickly.
  - They can be farmed on a large scale to produce useful complex molecules.
  - They can be **genetically engineered** to produce further valuable chemicals.
  - There are few ethical concerns about the way that a microorganism is treated.
  - Bacteria possess plasmids.

Product	Produced by	Use of product
Vitamin B12	*Streptomyces*	Vitamin supplements
Protease	*Bacillus subtilis*	Enzymes for biological washing powders
Pectinases	*Aspergillus* sp.	Increasing the amount of juice that can be extracted from fruit
Ethanol	*Saccharomyces* (yeast)	Making alcohol to use as a biofuel
Chymosin	*Escherichia coli*	An enzyme used to make vegetarian cheese from milk
Penicillin	*Penicillium chrysogenum*	A common antibiotic

Useful products made by microorganisms.

- Microorganisms such as bacteria can produce chemicals, such as proteins, very quickly.
- They are difficult to harvest because the cells are very small.
- People also associate bacteria with illnesses, rather than food.
- This is despite bacteria being involved in cheese and yoghurt making.
- Yeasts are easier to harvest, but grow more slowly and produce lower quality protein.
- Quorn™ is made of a single-celled protein produced by a fungus.
- It can be made to look like meat pieces or mince, so acts as a meat replacement for vegetarians.
- Fungi are also responsible for many of the antibiotics that we use, e.g. penicillin comes from the fungus *Penicillium chrysogenum*.
- The protein production rate from microorganisms is far more efficient than in large animals, such as cattle.

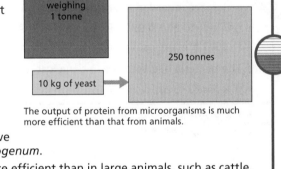

Output of protein in 24 hours

One cow weighing 1 tonne → 100 g

10 kg of yeast → 250 tonnes

The output of protein from microorganisms is much more efficient than that from animals.

Fermentation

- To produce large quantities of proteins and antibiotics, such as penicillin, a **fermenter** is used.

- Fermenters are a closed system through which all inputs and outputs can be carefully controlled and monitored.

- Inside the sealed vessel variables are controlled, such as:
  - temperature
  - pH
  - oxygen levels
  - nutrients.

- By manipulating the different inputs and outputs the manufacturers can produce large amounts of products rapidly and cheaply.

Improve your grade

Producing protein and penicillin with microorganisms

Higher: In the future it may be possible to grow bacteria that produce meat protein as a byproduct. Suggest why some people would disagree with this. *AO2* [3 marks]

Genetic modification

Changing genes

- Weeds can reduce the yield of a crop by between 35% and 100%.
- **Selective herbicides** kill certain plants, leaving others intact.
- Spraying a herbicide on a crop would kill the competing weeds.
- Over time the weeds evolved resistance to the herbicide.
- This means that the crop yield starts to decrease again.
- Pests also eat crops, reducing the yield.
- Spraying crops with pesticides causes not only the pest to die but other, innocent, insects.
- There are a number of diseases of crops caused by microorganisms such as fungi.
- Even though crops could be sprayed with antifungal sprays, the fungi start to evolve resistance.
- Genetic engineers can now find genes that give the crop a new characteristic that was not previously present.
- Crops can have genes added that make them produce chemicals that act as herbicides, pesticides and antifungal agents.
- Crops can also have genes added that make the plant resistant to newly developed herbicides and pesticides. These genetically engineered crops mean that the farmer can use new herbicides and pesticides on their crop.

- Insulin is used to treat patients with Type 1 diabetes.
- It used to be extracted from the pancreases of pigs or cows.
- The insulin had to be very carefully purified to prevent contaminants from the non-human species.
- There were ethical issues with animal insulin.
  - Cows are regarded as being sacred by certain religious groups.
  - Pigs are regarded as unclean by certain religious groups.
  - If you were a vegetarian or a vegan then you would be forced to use a product from an animal.
- In 1982 human insulin became available that had been obtained from genetically engineered bacteria.
- The bacteria had the human gene inserted and all the insulin produced was identical to human insulin.
- It is easier to produce the insulin in large quantities.

Recombinant DNA

- Genetic engineering of crops for herbicide resistance and human insulin in bacteria uses the same technique: recombinant DNA.
- **Recombinant DNA** combines DNA from more than one source.
- A **vector** (a plasmid or a virus) is used to transfer the gene from one organism to another.
- Plasmids are circular DNA molecules found in some bacteria.
- **Viruses** can inject their DNA directly into a cell.
- The desired gene, in the new organism, is now expressed when the modified cell develops.

desired gene is located in the source cell

isolated desired gene

enzymes used to multiply desired gene

plasmids multiplied

plasmid donor

vector

plasmid opened with restriction enzyme

isolation of bacterial plasmid

gene codes mixed with plasmids

some of the plasmids now contain desired gene

bacteria contain the plasmids

chromosome

bacteria containing desired gene selected and multiplied

bacteria containing plasmids inserted into plant cell

Recombinant DNA.

EXAM TIP

Be precise when discussing gene transfer. It is the desired gene from one organism that is removed and inserted into the target organism.

Improve your grade

Genetic modification

Higher: Explain how vectors are used in genetic engineering.

AO1 [4 marks]

Genetic testing

Testing the genome for a gene

- Scientists use a technique called Fluorescence *In Situ* Hybridisation (FISH) to find a particular gene in a person's cells.

- This technique is possible because the two strands of the DNA molecule are mirror images of each other.

- If you have one strand then only the matching opposite strand will match with it.

- The gene probe has the complementary matching DNA sequence for the gene that is being detected.

labeling with fluorescent dye

probe DNA that matches target gene

denature & hybridize

if correct sequence present, probe attaches

if gene present, fluorescence detected

Fluorescence *In Situ* Hybridisation (FISH).

- The process involves four steps.
  - Step 1: prepare a **DNA probe** – a length of DNA to bind onto the target gene.
  - Step 2: the DNA probe is built, with the correct DNA sequence and a fluorescent chemical attached.
  - Step 3: mix the DNA probe with white blood cells from the person who is being tested and leave for enough time for the probe and gene to match up.
  - Step 4: cells are viewed under the microscope and exposed to UV light. If the probe has matched with the desired gene, then the fluorescent chemical attached will glow.

- Before viewing the cells, the whole preparation is washed first to remove any unattached probes that would otherwise fluoresce with the UV light.

- The FISH technique is particularly useful for testing for the presence of genes that we know cause certain illnesses.

- It is used in chorionic villus sampling to check whether a fetus is carrying certain deleterious genes.

Remember!
Don't make the mistake of saying red blood cells. Red blood cells have no nucleus.

Probability and chance

- Evidence suggests that cancer is caused not by single genes but by a collection.

- Breast cancer for example has two main genes which, if not functioning correctly, increase the chance of getting cancer: BRCA1 and BRCA2.

- Just because a person carries a faulty gene it does not mean that they will get the disease.

- It does mean, however, that they will need to monitor their lifestyle (the environmental conditions) to ensure that their overall chance of getting the disease is reduced.

- Predicting possible illnesses creates some ethical issues.

- If the tests are done before birth, should the tested fetus be aborted before birth if found to be at risk?

- Who decides whether a particular disease or disorder is worth living with?

- What if the testing is abused so that certain idealistic traits are selected for, e.g. blue eye colour, etc.

- Even after birth, a genetic test could potentially affect someone's life chances, e.g. difficulty getting a job or insurance if you carried a gene that made it more likely that you will suffer from an illness.

Improve your grade

Genetic testing

Higher: Explain the steps involved with Fluorescence *In Situ* Hybridisation.

AO1 [4 marks]

Nanotechnology

The scale of nanotechnology

- **Nanotechnology** means technology that works on the scale of atoms and molecules.
- Materials at the nanoscale are usually between 0.1 and 100 nanometres (nm) in size.
- One nanometre (1 nm) is one billionth of a metre.
- It can also be written as 10^{-9} m, or 0.000000001 m.
- Most atoms are 0.1 to 0.2 nm wide.
- Strands of DNA are about 2 nm wide.
- Red blood cells are around 7000 nm in diameter.
- Human hairs are around 80 000 nm across.

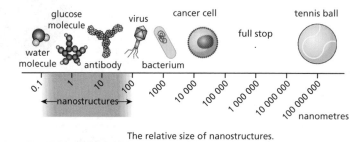

The relative size of nanostructures.

The uses of nanotechnology

- One potential use of nanotechnology is in preventing food waste.
- On average nearly 20% of food is wasted because it has gone bad.
- Most of the wastage is caused by the reaction of the foodstuffs with oxygen in the air.
- The oxygen reacts and changes the properties of the food, e.g. oxygen makes the cut surface of an apple turn brown.
- Oxygen is needed for the growth of microorganisms that can feed on the food.
- Traditional packaging cannot prevent the slow intake of oxygen.
- Using nanoparticles in the packaging makes it more difficult for the oxygen to enter.
- It also prevents water passing through in the other direction.
- Food is protected from spoiling for much longer, extending its shelf-life.
- Other nanoparticles can be added to packaging that change colour if the packing is damaged or the temperature inside has increased above a certain level.
- The change of colour warns people that the food is not safe to eat.
- Another class of nanoparticles can detect the presence of contaminants in food (such as those produced by bacteria and fungi) and change colour to show they are present.
- These technologies should ensure that people throw food out only when it has spoiled and not before.

- Nanoparticles can be used in medicine.
- Silver nanoparticles can be embedded in wound dressings and sticking plasters.
- When the silver nanoparticles are exposed to water they become antibacterial.
- This means that wounds are less likely to become septic.
- Gold nanoparticles can be used to attach to antibodies, which target cancer cells.
- When the antibodies bind to the cancer cells, infrared light passed into the body heats up the nanoparticles, destroying only the cancer cells.
- Nanoparticles used in this way can be a safer way to treat cancer as the nanoparticles are inert and the patient does not necessarily have to suffer the side effects from radiotherapy and chemotherapy.

Improve your grade

Nanotechnology

Higher: Suggest why a sticking plaster embedded with silver nanoparticles will be more effective than a pair of pyjamas embedded with silver nanoparticles. *AO2* [4 marks]

Future medicine

Rebuilding the heart

- The heart is an essential organ. Its role is to pump blood around the body.
- Anything that interferes with the function of the heart can have very serious implications.
- There are a variety of different ways that the heart can be damaged.
- One example is where the valves become damaged by disease.
- The valves prevent the backflow of blood. If damaged, this will not take place and can be life threatening.
- A solution is to transplant valves from a donor heart.
- Transplantation carries the risk of rejection by the immune system.
- Replacing the faulty valves with an artificial valve made of plastic has less chance of rejection because the blood does not stick to the plastic.
- Surgeons can also use heart valves from pigs specially bred to reduce the problem of rejection.
- There are ethical issues in using animal parts for human transplantation.
- The heart has specialised cells that control the heartbeat.
- If these become faulty then a pacemaker will be needed.
- A **pacemaker** is a mechanical device that monitors the heartbeat and, if it is not beating correctly, can give the heart a small shock to make it contract correctly.
- If the parts of the heart cannot be repaired or replaced then a whole human heart can be transplanted.
- Scientists are developing completely artificial mechanical hearts that remove the problems with rejection.

Using stem cells

- Currently if an organ becomes faulty you can either have it repaired or replaced with a donor or mechanical organ (if available).
- An alternative treatment could be using stem cells.
- **Stem cells** are cells that are undifferentiated.
- Given the correct biochemical signal they can develop into any type of cell in the body.
- There are different types of stem cells.
  - Embryonic stem cells – harvested from embryos (four to five days old).
  - Fetal stem cells – harvested from an aborted fetus.
  - Somatic stem cells – harvested from the sites in the body that naturally produce more cells. These are also known as adult stem cells.
  - Amniotic stem cells – stem cells harvested from the amniotic fluid surrounding the developing fetus.
  - IPS cells – these are stem cells formed by forcing an adult cell to revert back to being an embryonic stem cell.
- The use of stem cells is controversial. Some groups disagree with using stem cells that have come from an embryo or fetus. Some people are concerned about the threat of the stem cells becoming cancerous and being a greater threat than the disease they are trying to correct.
- Governments set up ethical committees to make the decisions needed as to whether a treatment should be allowed, or not.

- Stem cells can be used to treat **leukaemia** (a disease of the blood).
  - The body's normal stem cells that are not functioning correctly are destroyed by chemo- and radiotherapy.
  - Stem cells from a bone marrow donor are injected into the patient's bones.
  - They can then produce new blood cells.
- Stem cells are also being used to treat nerve cell damage.
  - When the spine is broken, normally the nerves do not regrow so paralysis is permanent.
  - Stem cells are being tested in trials to see whether they can reconnect the broken nerve tissue.

Improve your grade

Future medicine
Higher: Suggest why stem cells are being used to treat people who are paralysed. *AO1* [3 marks]

B7 Summary

The vertebrate internal skeleton supports the body, protects organs and enables movement.

When training or being treated for a skeletal-muscular injury, your lifestyle and family history need to be disclosed. Sometimes a physiotherapist will be the medical professional involved.

Joints can move through the antagonistic action of muscles.

Movement and exercise

Effectiveness of an exercise program can be monitored through looking at:
blood pressure
pulse rate
recovery time after exercise

Joints typically have: cartilage to reduce friction and ligaments to hold bones together, tendons to transmit force.

We use PRICE to treat a sprain.

BMI is calculated by:

$$BMI = \frac{body\ mass\ (kg)}{[height\ (m)]^2}$$

A double circulatory system means that the blood passes through the heart twice for every one circulation through the body.

The heart is a muscular pump. Blood enters via the vena cava into the right atrium into the right ventricle. It then moves into the pulmonary artery to the lung. It returns via the pulmonary vein into the left atrium and then the left ventricle, to be pushed around the body via the aorta.

Blood is a mixture comprising:
red blood cells (carrying oxygen)plasma (transporting gases, hormones, water, nutrients, waste products)
white blood cells (to fight infection) platelets (to clot the blood).

The heart and circulation

Valves in the heart and veins prevent the backflow of blood.

Red blood cells are adapted to their function by: having no nucleus, so they can be packed full of haemoglobin. They are also a biconcave shape to maximise the surface area.

Capillary beds enable the movement of dissolved gases and other chemicals into cells through the tissue fluid surrounding cells. Chemicals move down concentration gradients.

Temperature regulation enables the core body temperature to remain around 37°C.

Insulin is a hormone that controls blood sugar levels.

Temperature receptors in the skin and brain detect changes in temperature. If they are too high then the body will produce sweat, which cools the body down. Vasodilation causes blood to flow close to the surface of the skin.

Energy balance

There are two types of diabetes. Type 1 is inherited and means no insulin is produced. Insulin is secreted into the body by the pancreas. Type 2 diabetes is acquired, through poor diet and means insulin is no longer recognised by the body. Treatment is via diet and exercise.

If the body temperature gets too cold, the muscles can start to shiver (contracting rapidly) and the body respires more to produce heat. Vasoconstriction can restrict the blood flowing to the skin to reduce heat loss.

The hypothalamus controls temperature regulation. It processes information and coordinates a response using antagonistic effectors.

B7 Summary

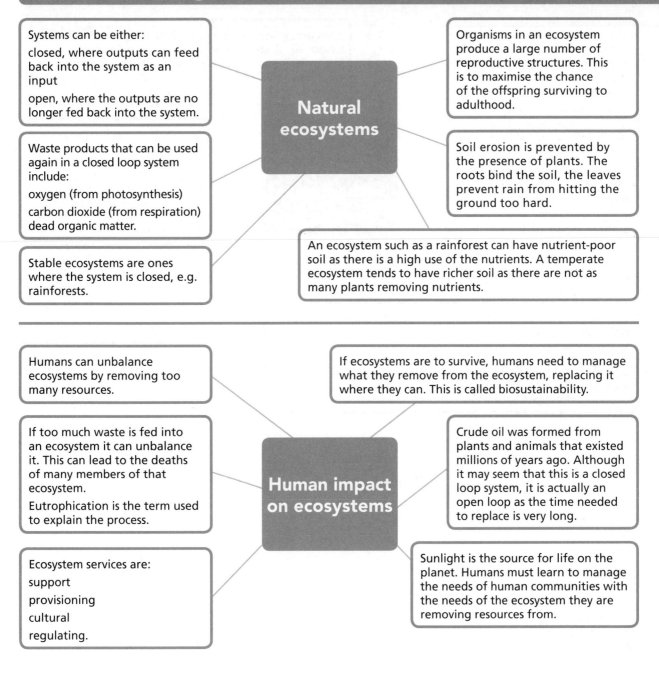

Natural ecosystems

Systems can be either:

closed, where outputs can feed back into the system as an input

open, where the outputs are no longer fed back into the system.

Waste products that can be used again in a closed loop system include:

oxygen (from photosynthesis)

carbon dioxide (from respiration) dead organic matter.

Stable ecosystems are ones where the system is closed, e.g. rainforests.

Organisms in an ecosystem produce a large number of reproductive structures. This is to maximise the chance of the offspring surviving to adulthood.

Soil erosion is prevented by the presence of plants. The roots bind the soil, the leaves prevent rain from hitting the ground too hard.

An ecosystem such as a rainforest can have nutrient-poor soil as there is a high use of the nutrients. A temperate ecosystem tends to have richer soil as there are not as many plants removing nutrients.

Human impact on ecosystems

Humans can unbalance ecosystems by removing too many resources.

If too much waste is fed into an ecosystem it can unbalance it. This can lead to the deaths of many members of that ecosystem.

Eutrophication is the term used to explain the process.

Ecosystem services are:

support

provisioning

cultural

regulating.

If ecosystems are to survive, humans need to manage what they remove from the ecosystem, replacing it where they can. This is called biosustainability.

Crude oil was formed from plants and animals that existed millions of years ago. Although it may seem that this is a closed loop system, it is actually an open loop as the time needed to replace is very long.

Sunlight is the source for life on the planet. Humans must learn to manage the needs of human communities with the needs of the ecosystem they are removing resources from.

DNA testing in humans involves extracting DNA from white blood cells. A gene probe, labelled with a fluorescent chemical, is added to the DNA to identify whether the gene or allele of interest is present. If it is, when UV light is shone on the sample, light will be produced.

Bacteria are used in industry because:

they reproduce rapidly

have plasmids that have a simple biochemistry

have fewer ethical issues.

New technologies

Nanotechnology is biology on the 0.1 to 100 nm scale. We can use nanotechnology to indicate when food is deteriorating, in food packaging to preserve the food for longer and to detect contaminants.

Bacteria can be used on a large scale to produce:

antibiotics and medicines single cell proteins

enzymes to replace animal enzymes.

Stem cells are cells which are undifferentiated. They can potentially be used to regrow cells that previously could not regrow, e.g. spinal nerve cells, brain cells.

Genetic modification, for example to produce insulin or herbicide resistance, involves:

isolating a desired gene from one organism

placing it into a vector (a plasmid or virus)

using the vector to transfer the gene to the desired cell and then detecting those cells which have the new gene inside.

Page 4 Studies on twins

Higher: How do studies of identical twins help us to understand the effect of the environment on the phenotype for a characteristic? *AO2* [3 marks]

Identical twins are born with identical sets of genes, so any differences in their phenotypes for a characteristic must be the result of an effect of their environment.

Answer grade: C. This answer is correct, but could be extended. It's important to spell out each stage of your answer, so the first part must refer to the genotypes of identical twins being the same.

For a B grade, you need to raise a more subtle point. Identical twins are usually brought up in a very similar environment, so studies of identical twins that have been separated at birth or at an early age are especially useful, as their environments may be very different.

Page 5 Variation in offspring

Foundation: Explain how variation occurs in the offspring produced by humans. *AO1* [5 marks]

Variation occurs because we all have different genes, which we inherit from our parents.

Answer grade: D. This answer is correct but lacks detail. For full marks at grade C, explain that it is because of *sexual reproduction* that we show variation. Be specific and say that we inherit half our genes from our mother and half from our father. However, siblings differ because they inherit a different *combination* of genes from their parents. Show how these combinations might occur, using diagrams of chromosomes and genes. You should also mention that the environment we live in also leads to variation in humans.

Page 6 Genetic diagrams

Higher: A disease called cystic fibrosis is caused when two recessive alleles of a gene are present. The diagram shows the occurrence of cystic fibrosis in the family.

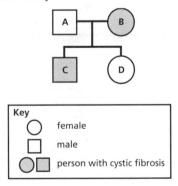

Key
○ female
□ male
◯▢ person with cystic fibrosis

What are the genotypes of:

a father A?

b daughter D?

Explain your answer. *AO3* [4 marks]

Father A is CC or Cc, because he does not have cystic fibrosis. As the daughter D is normal, she must be Cc.

Answer grade: B. Sentence 1 is partly correct, but it is possible to say with certainty what the genotype is. In sentence 2, the daughter's genotype given is correct, but the explanation is incomplete.

For full marks at grade A/A*, you need to *explain* that father A does not have cystic fibrosis, so must have at least one dominant gene. As the son has cystic fibrosis, the father must be Cc, and not CC (if he was CC, there would be no chance of producing a child with cystic fibrosis). The mother (B) has cystic fibrosis, so she must be cc. This means that daughter D must have inherited a c allele from her mother, but as she is normal, daughter D must be Cc.

Page 7 Gene disorders

Foundation: Write down the name of one dominant genetic disorder, and one recessive genetic disorder. For each disorder, list three symptoms the person will show. *AO1* [4 marks]

Huntington's disease: The symptoms are tremors, uncontrollable shaking and memory loss.

Cystic fibrosis: The symptoms are thick gluey mucus affecting the lungs and difficulty breathing.

Answer grade: E. In the question, there are 2 marks for naming the disorders and 1 mark for each list of three symptoms. The student has not said which is the dominant genetic disorder (Huntington's disease) and which is the recessive (cystic fibrosis). These are required to get the 2 marks.

For full marks, you must write down **three** symptoms for each disorder, as stated in the question, to get one mark for each. For Huntington's disease, these also include an inability to concentrate and mood changes. Note that 'tremors' and 'uncontrollable shaking' are the same thing, so three symptoms weren't given in the answer. For cystic fibrosis, symptoms also include chest infections and difficulty in digesting food.

Page 8 Stem cells

Higher: Discuss the future use of stem cells in medicine. *AO1, AO2* [4 marks]

Stem cells are unspecialised, so they can be used to produce different types of body cells. So stem cells can be used to renew damaged or destroyed cells, for example in spinal injuries, heart disease, Alzheimer's disease and Parkinson's disease.

Answer grade: C. This answer is correct; the student has referred to the fact that stem cells are unspecialised, so have the potential to develop into different cell types. However, the student has just focused on the possible replacement of damaged or destroyed cells.

For a B grade, it's important to mention other uses too. Stem cells can be used to improve our understanding of how cells become specialised. This occurs in the early stages of a person's development, by the switching on and off of particular genes. Stem cells can also be used in the testing of new drugs.

Page 10 Our defence system

Higher: A bacterium enters your blood stream. Describe the series of events leading to the bacterium being destroyed by your immune system.　　*AO1* [5 marks]

Antibodies are produced. The antibodies lead to the destruction of the bacterium.

Answer grade: B. The answer correctly states that it is the production of antibodies that leads to the bacterium being killed, but it does not describe the series of events that lead to its destruction.

For full marks, explain that the antibodies are made by white blood cells and how this is done. You could explain that the invading bacterium (antigen) becomes attached to a white blood cell with a matching antibody; the white blood cell then divides and the white blood cells all make many antibody molecules, which attach to the invading bacterium and destroy it.

Page 11 Vaccination programmes

Higher: The graph shows the number of cases of measles, and deaths from measles, in England and Wales from 1940 to 2006.

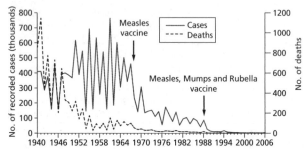

Discuss what the data suggest about the effectiveness of the measles vaccines.　　*AO3* [4 marks]

The graph shows how the introduction of the measles vaccine reduced measles cases. When the combined measles, mumps and rubella vaccine was introduced, it reduced these further.

Answer grade: C. The answer says, correctly, that the vaccines reduced the cases of measles, but it lacks detail, and doesn't describe how the *deaths* from measles changed over the time period.

For full marks, you would need to say that the numbers of cases of measles showed fluctuations (some very large) across this time period (and even after the introduction of the vaccines). The *deaths* from measles have shown a continuous decrease, even *before* the vaccines were introduced. You could also comment on the small peak around 1994. Remember, the data may show more than one trend, so you may need to describe and explain these one by one.

Page 12 Trialling new treatments

Foundation: When testing a chemical on humans to see if it would be suitable as a new antibiotic, a placebo is sometimes used.

Explain why placebos are important, and how ethical issues with using placebos are overcome. *AO1* [4 marks]

Placebos are important because patients often feel an improvement even if the treatment does not have the active component. Ethical problems arise if the group provided with the new treatment begins to improve.

Answer grade: D. The answer says why placebos are important, but does not explain fully what a placebo is. An ethical problem with the use of placebos is given but there is no indication as to how this problem is overcome.

For full marks, explain that a placebo is a tablet or solution made to look just like the new drug, but without the active ingredient. Patients sometimes show an improvement when given any kind of 'treatment', so the use of a placebo will suggest if this is the case, and help to distinguish this from improvement resulting from the 'real' treatment. The ethical problem given can be overcome if members of the group using the placebo are switched to the treatment so they can benefit from it also.

Page 13 The heart and circulatory system

Foundation: The heart needs its own blood supply to live. Describe how the heart receives this and what happens in a person with coronary heart disease.　　*AO1* [5 marks]

The coronary arteries supply the heart with blood. In coronary heart disease, the coronary arteries become blocked with fatty deposits. The person has a heart attack.

Answer grade: D. This answer begins by stating that the coronary arteries are involved, but there is no mention of how these keep the *whole* of the heart supplied with oxygen (they run over the surface of the heart, so all the heart receives blood). Also, while the answer states that the coronary arteries become blocked and heart attack results, there is no description of how this happens. You should explain that, when the coronary arteries get blocked, the heart is deprived of oxygen (or part of it will be), so the heart muscle will die.

Page 14 Heart rate and blood pressure

Foundation: Explain the term 'blood pressure' and describe how blood pressure affects health.

　　AO1 [2 marks]

Blood needs to be under high pressure to be pumped around the body, otherwise it would reach all parts of the body. High blood pressure increases the chance of heart attack and stroke.

Answer grade: E. The first part clearly explains why blood needs to be under high pressure to be pumped around the body. The next sentence states the effects of high blood pressure, but does not mention the effects of low blood pressure. Low blood pressure will result in dizziness and fainting.

Page 15 The kidneys

Higher: Explain the effect of Ecstasy (MDMA) on the water balance of a person's body.　　*AO1* [3 marks]

Ecstasy increases the production of anti-diuretic hormone (ADH) by our pituitary gland. So Ecstasy causes our body to retain water.

Answer grade: B. This answer *describes* the effects of Ecstasy rather than *explaining* them. To get full marks you would need to explain that by increasing the amount of ADH – a hormone that prevents urine production – Ecstasy reduces the amount of urine produced. You could use a diagram to illustrate this.

Page 17 Extinction

Foundation: The harlequin ladybird originated in Asia and arrived in Britain in 2004. It eats the same food as our native ladybird (aphids – greenfly and blackfly). It also predates on many insects (other ladybirds, and the eggs of butterflies and moths) and eats fruit.

Suggest why the spread of the harlequin ladybird might affect food webs. *AO2* [4 marks]

As it eats the same food as our native ladybirds, it might out-compete them for food. Our native ladybirds will die. As it is a predator, it will prey on many other species of insect. It may also eat fruit.

Answer grade: D. The answer correctly refers to competition for food and predation of native species, and ladybirds eating fruit, but does not explain how these affect food chains. To gain full marks, you would need to explain that invasive species can also bring disease (affecting native ladybirds and perhaps other insects). By eating fruit the harlequin ladybird may deprive other animals of food, and this may affect the reproduction of the plants.

Page 18 Indicators of environmental change

Foundation: Lichens are sensitive to sulfur dioxide in the air. The distribution of three different types of lichen was measured in a city centre, and at different distances from it. The results are shown below.

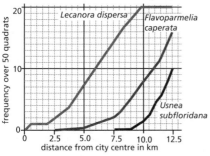

a Suggest what the graph tells you about the resistance of the lichens to pollution. *AO3* [1 mark]

b Why did the scientists measure the frequency of the lichens in 50 quadrats? *AO2* [2 marks]

a The data suggest that Lecanora dispersa is the most resistant to pollution.

b To improve the reliability of the results.

Answer grade: E. The answer to part **a** tells us that *Lecanora dispersa* is the most resistant to pollution. To get full marks, put the lichens into order. *Lecanora dispersa* is the most resistant, followed by *Flavoparmelia caperata*, with *Usnea subfloridana* being the least resistant.

Part **b** states, correctly, that taking counts from 50 quadrats will provide reliable results, but this answer is vague and needs to be explained further. You need to say that taking counts from a larger number of quadrats (than say, 10) is better, as using 10 quadrats, or fewer, may not truly reflect how the lichen is distributed. This will increase the confidence you have in the results.

Page 19 Variation and natural selection

Higher: Scientists have observed that the number of cases in which disease-causing bacteria are resistant to antibiotics has increased over the last 30 years. Explain why some people think that this is evidence of natural selection. *AO2* [4 marks]

The bacteria have genes that makes them resistant to antibiotics. This is evidence for natural selection because only the bacteria that have these genes are able to survive and reproduce.

Answer grade: B. While these points are correct, the answer lacks detail and does not link the points together or describe the sequence of events. For full marks, you should say that in any population of bacteria there is natural variation. When exposed to the antibiotic, most of the bacteria will be killed, but some may survive, because they have genes that make them resistant. These bacteria will reproduce, so these 'resistant genes' will spread throughout the population. The bacteria that survive have therefore been *selected* by the antibiotic, and over a short time, the species of bacterium will become resistant to the antibiotic. You could illustrate this process nicely with a diagram.

Page 20 Evidence for evolution

Higher: The following are sequences of very short sections of DNA from four species of primate – a human, Neanderthal man, a chimpanzee and a gorilla.

Human	**CTGGGCGCGTGCGGTTGTCCTGGTCCTGCT**
Neanderthal	**CCGGGCGCGAGCGGTTGTCCTGGTCCTGCA**
Chimpanzee	**CCGGGCGCGTGCGGTTCACCAGGTCCTGCA**
Gorilla	**CAGGGCGCGGGAGGTTTACCACATGCTTCA**

Look at the sequences and suggest what the evolutionary relationships of the animals are based on the data. Explain the level of confidence you have in your conclusion, and suggest how this could be increased. *AO2, AO3* [5 marks]

The Neanderthal is closest to humans, followed by the chimpanzee, then the gorilla.

The comparison is of only 30 bases, of the millions found in each organism, so many more need to be analysed.

Answer grade: B. This answer is correct, but the reasoning in the first sentence could be quantified, and the second sentence makes just one point.

For full marks, explain that the sequences of the human and Neanderthal have three differences, the human and the chimpanzee have five differences, and the human and the gorilla have eleven differences. Explain that the similarities not only indicate the closeness, but also how long ago the branches may have occurred. You could extend the comparison in the second sentence by looking at other evidence, such as that from the fossil record, and from similarities in other patterns of the organisms, e.g. their skeletons.

Page 21 Conserving biodiversity

Foundation: When investigating species extinctions, explain why we can use real data of human populations but need a computer model for numbers of extinctions. *AO2* [4 marks]

It is possible to record the numbers of the human population, but not species of organisms.

Answer grade: D. The answer is correct, but to get full marks you should say that the actual number of species of organisms on the planet is unknown – many are still undiscovered. It's not possible to record the extinction of a species if we don't know that it exists, so it's only possible to estimate these. And of course we cannot do this for the past.

Page 59 Enzymes

Higher: Explain what happens to enzyme activity as the temperature is increased.   *AO1* [5 marks]

As the temperature increases, enzyme activity increases up to a certain temperature. In most enzymes, activity begins to decrease above about 40 °C, and ceases at about 60 °C, because the shape of the enzyme is changed.

Answer grade: B/A. This answer gains marks for describing the pattern of enzyme activity correctly and accurately stating that this pattern is shown by most enzymes (some enzymes found in bacteria can work well between 80 and 100 °C). The answer does not give reasons why enzyme activity increases, however, which is owing to an increase in the reaction rate at higher temperatures. Furthermore, the answer says that enzyme activity stops at around 60 °C, but does not explain why.

To gain full marks you would need to explain that enzyme activity stops at around 60 °C because, at this temperature, the heat changes the shape of the active site (by breaking bonds in the enzyme molecule), so the substrate will no longer fit into it. At this point, the enzyme is said to be denatured.

Page 60 Glucose: making it and using it

Foundation: Explain how the products of photosynthesis are used by the plant.
          AO1, AO2 [5 marks]

The products of photosynthesis are used in respiration. They are also used to produce the chemicals required for growth.

Answer grade: D/C. The first sentence gains 1 mark but does not mention that glucose is the main product of photosynthesis that is used for respiration. The second sentence also gains 1 mark, but lacks detail. For full marks, you need to discuss the chemicals used for growth that glucose is used to produce. These include starch for storage in cells and structures such as seed, cellulose for cells walls, and proteins for growth.

Page 61 Moving chemicals in and out of plants by osmosis

Higher: Explain how plant roots take up water, and how this water moves across a plant root.   *AO2* [5 marks]

The plant roots take up water by osmosis, because water is in a higher concentration in the soil than in a root cell. This dilutes the water in the first plant root cell, so water moves across the root.

Answer grade: C/B. The first sentence is correct, but could have begun with a definition of osmosis, for example: 'Osmosis is the net water movement from an area where it is low concentration to where it is in high concentration.'

The second sentence is also correct, but does not explain fully why water should move. For full marks, you would need to describe how movement of water into the first root (hair) cell dilutes the cell contents, so the water concentration is now higher than in the cell next to it, deeper into the root. Water will move into this second cell by osmosis, whose water concentration is now increased, so water will move into the third cell, and so on. This process could be illustrated in an annotated diagram.

Page 62 Investigating the effects of light on plant growth

Foundation: Describe and explain how an ecologist would compare how a plant is distributed in two meadows.   *AO1, AO2* [5 marks]

Throw a quadrat ten times in the first meadow, then in the second, and count the numbers of the plant in each quadrat. For each meadow, calculate the mean number of plants per quadrat.

Answer grade: D/C. The answer is correct, but the description lacks some important detail. For full marks, you would need to suggest an appropriate size of quadrat. For example, for a meadow an appropriate area of quadrat would be 1 x 1 m or 0.25 x 0.25 m, depending on the size of the meadow. It should also be emphasised that the quadrat should be thrown or placed (using coordinates) at random. Though not essential, for effective comparison it is best to calculate the mean number of plants per metre square in each meadow.

Page 63 Fermentation

Foundation: Explain what the graph below tells us about changes in concentration of sugar and ethanol as yeast grows.

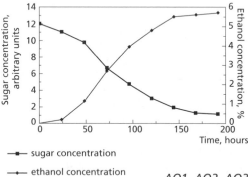

—■— sugar concentration

—◆— ethanol concentration   *AO1, AO2, AO3* [5 marks]

Yeast is feeding on the sugar, which decreases in concentration. It is producing ethanol by the process of fermentation, so its concentration is increasing.

Answer grade: C/B. Sentence 1 is correct, though it is better to say that the yeast is using the sugar for respiration. Sentence 2 is also correct but lacks detail. It does not mention that yeast produces ethanol under anaerobic conditions. The rise in ethanol concentration for the first 24–25 hours is slow. This is because, at this stage, anaerobic respiration is taking place only very slowly (oxygen is available to the yeast but later runs out).

For full marks you would also need to explain that the graph shows that ethanol concentration (and sugar concentration) eventually levels off, because the ethanol is inhibiting the growth of the yeast and/or the sugar is running out.

Page 65 Cell specialisation in animals

Foundation: Compare how cells become specialised in animals and plants. *AO2* [5 marks]

In animal cells, specialised cells are produced from stem cells. In plants, they are produced following cell division in regions called meristems.

Answer grade: D/C. Both of these statements are correct, but the student does not go on to add detail. For full marks, you should mention embryonic and adult stem cells in animals. You need to explain that cells become specialised in the embryo after the eight-cell stage, when cells produced by embryonic stem cells differentiate. You should also say that some stem cells remain in the adult (adult stem cells), and that these can differentiate into a limited number of cell types. Finally, you need to state that in plants, when meristem cells divide into two, one of the new cells produced by the meristem can differentiate.

Page 66 Plant clones

Foundation: Explain how and why plant breeders who have produced a new variety of plant take many cuttings from it. *AO1, AO2* [5 marks]

Plant cuttings are taken from the new variety by cutting off a shoot from the plant and placing it in compost that is kept moist (or sometimes water). After around two weeks, the cutting will produce roots, and will grow into a new plant. Plants grown from cuttings are identical to the parent plant.

Answer grade: D/C. Sentence 1 is correct, but does not mention that the cutting is usually dipped in hormone rooting powder to help it to root. The final paragraph is also accurate, in that the plants are identical to the parent, but it does not relate this to the question, i.e. its importance when producing many identical copies of the new variety of plant.

For full marks, you need to say that the root cutting is dipped in hormone rooting powder, in order to promote the growth of roots from the portion of the cutting under the surface of the compost. You also need to say that as the plants grown from cuttings are identical to the parents, the characteristics of the new variety will be present in all the plants produced.

Page 67 Mitosis and meiosis

Higher: Explain why gametes (sex cells) are produced by meiosis and not by mitosis. *AO1, AO2* [4 marks]

Meiosis is used to produce gametes in order to keep the chromosome number constant (46 in humans) from generation to generation.

Answer grade: C/B. The sentence is correct, but offers little in the way of explanation, or comparison with mitosis, so only gains 1 of the 4 available marks.

For full marks, you first need to say that human cells contain 46 chromosomes, which can be assembled as 23 pairs (each one of the pair carrying the same type of genes). If the cells that produce gametes divided by mitosis, at fertilisation, the zygote/offspring would have 92 chromosomes, so the chromosome number would double every generation. These cells therefore divide by a process called meiosis, so each gamete in humans contains 23 chromosomes; one from each pair. This means that the chromosome number is restored on fertilisation, with the zygote having 23 pairs of chromosomes.

Page 68 Protein synthesis

Higher: Describe the process by which proteins are produced. *AO1* [5 marks]

Proteins are assembled on a ribosome. The order of the amino acids in the protein is specified by the genetic code of the DNA.

Answer grade: B. Both of these statements are correct, but the description lacks some important detail.

For full marks, you should be begin by saying that in the nucleus of the cell, messenger RNA is synthesised using the DNA of the gene (the gene that codes for this protein) as a template. The mRNA passes into the cytoplasm and attaches to a ribosome. The amino acids are ferried in to the ribosome and the amino acids are bonded together (in the order specified by the genetic code).

Page 69 Stem cell research and therapy

Higher: Scientists have reprogrammed skin cells to function as stem cells. Explain, in principle, how this technique is carried out, and why this might be preferable to using embryonic stem cells. *AO2* [5 marks]

Skin cells have been changed into stem cells using a chemical treatment. This method is preferable to using embryonic stem cells, as when these are removed from the embryo, the embryo is destroyed.

Answer grade: C/B. Sentence 1 is correct, although it does not explain what the chemical treatment does. Sentence 2 is also correct, and points out the main ethical problem with using embryonic stem cells, but does not point out the deficiencies of using the transformed cells.

For full marks at grade A, you need to say that the chemical treatment reactivates genes that have become inactive, so that the transformed cells can develop into different cell types. You also need to explain that these treatments are in the early stages of their development, and although a number of cell types have been produced, cells have not yet been produced that will develop into all cell types.

Page 71 Neurons

Higher: Explain how nerve cells (neurons) are adapted to transmitting nerve impulses. *AO2 [5 marks]*

Nerves are the longest cells in the body as they have to reach all parts of the body. They have a long extension to the cell called the axon. The axon is insulated by a fatty covering called the myelin sheath.

Answer grade: D/C. Both of these statements are correct, but the student has missed some important points.

For full marks, begin by saying that the nerve impulse is an electrical impulse (which explains why it's important to be insulated). To extend the answer to an A grade, say that the presence of the myelin sheath not only insulates the neuron, but also enables much greater transmission speeds, as the nerve impulse jumps from one gap in the sheath to the next.

For a C grade it's important to mention that the neuron has extensions called dendrites, which enable it to communicate with other neurons. A more subtle point, at A/A* grade, is that the end of the axon contains chemical transmitter molecules that enable it to communicate with other nerve cells and other effectors.

Page 72 Synapses

Higher: Describe how a nerve impulse is transmitted from a sensory nerve to a nerve close to it in a spinal cord. *AO1, AO2 [5 marks]*

As the nerve impulse reaches the end of the nerve, a chemical transmitter is released. This passes across the synapse, and sets up a nerve impulse in the nerve on the spinal cord.

Answer grade: B. The answer is correct but misses some detail. Also, although the student has said correctly that the nerve passes across a synapse, they have not defined what a synapse is.

For full marks, you need to say that nerves are not connected together physically; instead a chemical transmitter is released from the first nerve and passes across a gap called a synapse. Point out that the type of chemical transmitter used is dependent on the location and type of nerve. Finally, you need to describe how, after the impulse has passed, the remaining chemical transmitter in the synapse is reabsorbed into the first nerve (or alternatively broken down by an enzyme).

Page 73 Instinctive and learned behaviour

Higher: A bird eats a poisonous, brightly coloured caterpillar. It is sick, but survives. In future, it avoids eating this type of caterpillar. Explain how this is an example of a conditioned reflex and how it might help the bird's survival. *AO2 [5 marks]*

After being sick, the bird learned to avoid the poisonous, brightly coloured caterpillars. It had associated the bright colours of the caterpillar with the unpleasant experience. This is called a conditioned reflex.

Answer grade: C. While this answer is correct, the student has not defined the two stimuli involved. For full marks, you need to define the poisonous/distasteful nature of the caterpillar as the primary stimulus, and the bright colours of the caterpillar as the secondary stimulus.

The student has also not explained how the response involved in the conditioned reflex – avoiding brightly coloured caterpillars – has no direct connection with their distastefulness or poisonous nature. You need to explain that, after tasting the caterpillars once or possibly a few times, the bird would come to associate the bright colours with distastefulness.

Finally, you need to mention how this can help the bird's (and the caterpillar's) survival. In being sick, the bird removed the poisonous caterpillar from its gut, but on another occasion may have eaten sufficient or kept it in its gut for long enough to kill it. So in not eating the caterpillar again, poisoning would be avoided.

Page 74 Brain structure

Foundation: Describe how scientists have mapped the areas of the brain to see how it works. *AO1 [3 marks]*

Neuroscientists have studied people with brain injuries and investigated how people react when their brains are stimulated using electrodes.

Answer grade: D. Both of these statements are correct, but the answer lacks detail. For full marks, you should refer to invasive and non-invasive techniques, and describe these. It's also important to say how the effects of brain injury are studied.

The answer also gives no information on non-invasive techniques, e.g. scanning techniques such as MRI scanning. You need to explain that these are used to compare the structure and activity of the brains of healthy people and people with brain disease, and when a person is stimulated by music, language, etc.

Page 75 Drugs

Foundation: Some chemicals affect how nerve impulses are transmitted across synapses. Give **two** examples of these chemicals, and state how these chemicals work. *AO1, AO2 [3 marks]*

Prozac increases levels of a chemical transmitter substance that carries the impulse between nerves. Toxins can block certain chemical transmitters.

Answer grade: E. The first sentence is correct, and is complete, as the question only says 'state' and doesn't ask for a description. The second sentence is also correct, but does not give an example, just a type of chemical that affects transmission. For full marks, you need to provide an example of a toxin that blocks a chemical transmitter, e.g. curare, which is a poison used on the tips of arrows by South American Indians, or botulin toxin ('botox').

Page 41 How the body moves

Higher: If someone tears their Achilles tendon (the tendon connecting the heel to the calf muscle at the back of the leg), they will be unable to stand on their toes. Explain why this is the case. *AO1* [4 marks]

If the tendon is torn then the person couldn't stand on their toes because the tendon is not attached. The foot would not be able to point downwards. This is because tendon attaches muscle to bone.

Answer grade: C. This answer is correct, but needs to be extended. The answer has stated what tendons do but not why the person cannot stand on their toes.

For a B grade you need to mention that the Achilles tendon is attached to an antagonistic muscle, which is one of an antagonistic pair. When it contracts the foot is pulled down. If the tendon is cut then the foot will only be able to be lifted up.

Page 42 Joints

Foundation: Label a diagram of a knee joint. *AO1* [2 marks]

Example answer:

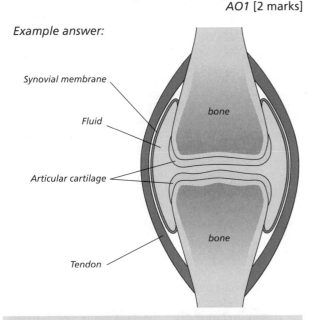

Synovial membrane

Fluid

bone

Articular cartilage

bone

Tendon

Answer grade: D. This answer has the ligament mislabelled as a tendon and, although the fluid is correct, it is not specific enough. For full marks at grade C correctly label the ligaments and the synovial fluid and make sure that fluid is correctly specified as synovial fluid.

Page 43 Exercise and health

Foundation: Why is BMI used to decide whether someone is overweight or not? *AO1* [3 marks]

BMI uses your body mass divided by the height^2 to work out a number. If the number is too high then you may be overweight.

Answer grade: D. The answer is correct, but not detailed enough to score full marks. For a C grade the equation for BMI should be given. This makes it easier to then explain why your BMI gets higher the more mass you have. The higher the mass for a given height, the higher the BMI number. If the BMI is above a set value, then it could indicate that the person is overweight, or not.

Page 44 Exercising safely

Foundation: Three competitors take part in a marathon. Arwen, Roz and Peter are aged 32, 45 and 65 respectively. Show by calculation their respective maximum heart rates and give the highest heart rate they should exercise at. *AO2* [3 marks]

Arwen = 188, Roz = 175, Peter = 155.

Maximum rate for Arwen = 169, Roz = 158, Peter = 150

Answer grade: D. The answer is almost fully correct. Unfortunately the maximum heart rate calculated for Peter is incorrect. To achieve a C grade it would be a good idea to show the working out for each (e.g. 220−32 for Arwen and 0.9 × 188 for the highest heart rate).

Page 45 How the heart works

Higher: Suggest why patients in a persistent vegetative state have to be turned regularly in their beds. *AO2* [4 marks]

Patients have to be turned regularly to make sure they don't get bed sores.

Answer grade: D. The answer here is correct, but does not explain why the patient needs to be turned. There are 4 marks available and you need to give a reason for the points made. To achieve a B grade you need to say that when a patient is lying down and not turning, the valves in the veins do not work as well at preventing backflow of blood. The blood can accumulate and become infected. By turning the body the blood is able to move.

Page 46 Blood components

Higher: Lorna is in training for the Rio Olympics. She trains at a high altitude, which increases the red blood cell content of her blood. Explain how this may give her an advantage over other competitors. *AO2* [6 marks]

Lorna trains at high altitude because this makes more red blood cells. She needs red blood cells when running, for respiration.

Answer grade: C. The answer has made the connection between red blood cells and respiration. More detail is needed to achieve a higher grade. For an A grade you need to link the need for more blood cells at high altitude, as there is less oxygen available. When back at low altitude the blood will still be higher in red blood cell count so Lorna's body may carry more oxygen than a competitor. The effect will wear off over time, however, and so the advantage is only temporary.

Page 47 Blood as a transport system

Foundation: Mark wants to have his DNA tested. Which part of his blood will be used to test his DNA? Explain your answer. *AO1 [3 marks]*

Mark will use his white blood cells to test his DNA.

Answer grade: D. The answer correctly identifies that white blood cells are needed for DNA tests. It does not explain why however, and loses two marks. To get a C grade the answer needs to add 'because the white blood cells have a nucleus packed full of DNA and the red blood cells do not have a nucleus or DNA'.

Page 48 Keeping cool

Higher: Niamh is suffering from heat exhaustion. Her friend Amy thinks that she is probably alright because she is not even sweating. Give reasons why Amy is making an incorrect diagnosis. *AO2 [4 marks]*

If someone sweats then their body is trying to cool itself down. As the sweat evaporates it takes some of the heat away, cooling the body down.

Answer grade: C. The answer has only explained how sweating helps, it has not addressed why Niamh has stopped sweating and the reason why this is dangerous. To achieve a B grade you need to explain that if sweating has been going on too long the body stops the process to conserve water. Niamh may have stopped sweating because of this. She needs medical attention.

Page 49 Keeping warm

Higher: Explain the role of the hypothalamus in temperature regulation. *AO1 [4 marks]*

The hypothalamus measures the temperature of the blood and sends messages back to the effectors to carry out a response.

Answer grade: C. The answer is correct but is vague. More detail is needed to show that you understand what the hypothalamus does. To achieve full marks for this question you would need to say that the hypothalamus is located in the brain and measures the temperature of the blood flowing through it. It coordinates antagonistic receptors, such as shivering and sweating, to make sure that the temperature is maintained.

Page 50 Diabetes

Higher: Explain why someone who has Type 2 diabetes does not always / necessarily need to inject insulin, whereas a person with Type 1 diabetes does. *AO1 [4 marks]*

A person with Type 2 diabetes doesn't need to inject insulin because they control their diabetes by changing their diet.

Answer grade: C. The answer correctly gives the method for blood sugar control in a Type 2 diabetes patient. However the question also mentions a Type 1 diabetes sufferer, so you should also be writing about them. To achieve an A grade you would need to state that in Type 1 diabetes insulin is not made by the body, so it has to be injected. With Type 2 diabetes, the insulin is still produced but it is not recognised, so injecting insulin would be pointless. Instead a Type 2 sufferer has to alter their diet to control sugar intake.

Page 51 Cycles in nature

Foundation: Explain why the manufacture of paper is a closed loop process. *AO1 [2 marks]*

Paper is made from trees. Once grown the trees can be replanted. This is a closed loop process because it's a cycle.

Answer grade: E. The answer has not defined what is meant by a closed loop process. It has indicated that trees can be replanted, which suggests a cycle, but this needs to be made clearer. To achieve a C grade you would need to say that a closed loop system is where the outputs at the end of the process are used as inputs at the start. Paper is made from trees, used, then pulped, and then used to make paper again.

Page 52 Cycles in rainforests

Foundation: Frogs produce a large number of frogspawn. Suggest why they do this, rather than produce one or two offspring. *AO2 [4 marks]*

Frogs don't give birth to live young, they make frogspawn instead. The spawn has to be in water. As the frogs don't look after the spawn a lot needs to be made, unlike humans who look after their young.

Answer grade: D. The answer is partially correct. The key part of the answer given is with the comparison between humans looking after their young and frogs. A better answer, which could achieve a C grade, would focus on the survivability of the frogspawn itself. Other organisms might eat it, or poor weather may cause a large number not to be successful. Hence large numbers have to be produced, unlike humans who protect and look after their young.

Page 53 Protecting soil

Foundation: What are the four main components of soil? *AO1* [3 marks]

The four components of soil are: minerals, worms, air and water.

Answer grade: D. The answer is mostly correct. Worms, however, is not accepted as meaning organic material. To achieve a C grade you need to include air spaces and organic matter in the answer.

Page 54 Ecological services

Higher: Explain why soil in the UK may be better quality than the soil in the Amazon rainforest. *AO1* [4 marks]

In the UK the weather is not as reliable as in the Amazon, which is on the equator. The soil in the UK will have more nutrients in it as there are not as many plants living in it.

Answer grade C. This answer is nearly there. It is lacking the reason why the soil in the Amazon is of a lower quality. With comparison questions always try to speak on both examples from the question. To get an A grade you would need to say that as plants do not grow as well in the temperate UK they do not remove as many nutrients from the soil. As the cycling is slow, the soil is richer. In the Amazon the plants grow rapidly and take the nutrients from the soil quickly. Cycling of nutrients is fast and the soil is consequently poorer.

Page 55 Poisoned lakes

Higher: Explain the term eutrophication. *AO1* [3 marks]

Eutrophication is where chemicals enter rivers and streams and cause the plants to grow and die. Eventually the whole river is unable to support life.

Answer grade C. The answer given is correct, but due to it not being specific it does not get a higher grade. To achieve a B grade you need to make sure that you say that the chemicals contain nitrogen compounds that cause increased plant and bacterial growth. The bacteria use up the available oxygen, killing larger organisms. The dead organisms rot and cause the oxygen levels to drop further still.

Page 56 Sustainable fishing

Higher: A new salmon farm is proposed for a lake in the Lake District. Suggest a biological reason why some people may protest against it. *AO2* [4 marks]

The salmon farm would probably not look very good and spoil the view. It would also disrupt other organisms in the lake.

Answer grade D. The first part of the answer is irrelevant, the question is asking for a biological reason. The second part just manages to get a mark for linking the farm with an effect on other organisms in the lake. To achieve a higher grade in this question you need to explain that the salmon are fed special food, which will probably leave the farm and enter the water, feeding other organisms in the lake. This may disrupt the food chain in the lake.

Page 57 Producing protein and penicillin with microorganisms

Higher: In the future it may be possible to grow bacteria that produce meat protein as a by-product. Suggest why some people would disagree with this. *AO2* [3 marks]

Some people are vegetarians. The gene the bacteria are using may have come from an animal and so the bacterial meat protein would still not be edible.

Answer grade C. The answer has made an interesting point but is missing the main issue, which is the possibility of disease. To achieve a B grade you would need to say that, for many people the idea of eating meat created by bacteria is unacceptable as they associate bacteria with diseases. This is despite a number of foods being made using bacteria.

Page 58 Genetic modification

Higher: Explain how vectors are used in genetic engineering. *AO1* [4 marks]

Vectors such as plasmids have a gene added. They multiply and are then put into the target organism. The plasmid carries the gene, which now works in the new cell.

Answer grade B. The answer discusses one of the two vectors but doesn't clearly mention the source of the gene. To achieve an A grade you would need to say that a desired gene is located in another organism. The desired gene is removed and added to a vector (a virus or a plasmid). Once there are enough copies of the vector it is introduced to the target organism. The gene is now in the target organism and will hopefully express the desired product.

Page 59 Genetic testing

Higher: Explain the steps involved with Fluorescence *In Situ* Hybridisation. *AO1* [4 marks]

A DNA probe is made. A fluorescent chemical is attached. It is then mixed with blood cells from the person who is being tested and left for a time. The cells are looked at under a microscope – if they fluoresce then the gene that is being tested for must be present.

Answer grade B. This answer is almost completely correct. The main error is in mentioning blood cells instead of white blood cells. To achieve full marks for this question you need to specify white blood cells. Remember, red blood cells don't have nuclei and saying 'blood cells' is therefore too vague.

Page 60 Nanotechnology

Higher: Suggest why a sticking plaster embedded with silver nanoparticles will be more effective than a pair of pyjamas embedded with silver nanoparticles. *AO2* [4 marks]

Plasters are in contact with wounds and so the silver nanoparticles will work whereas the pyjamas are not in contact with wounds so it is a waste.

Answer grade C. This answer is correct but has missed out some vital explanation of why the silver nanoparticles in a dressing work. To get an A grade the answer needs to mention that silver nanoparticles only work when they come into contact with water, which they will with a sticking plaster but not with a pair of pyjamas (other than when the skin sweats).

Page 61 Future medicine

Higher: Suggest why stem cells are being used to treat people who are paralysed. *AO1* [3 marks]

When you are paralysed, you cannot work because the nerves are broken. Stem cells might regrow the nerves.

Answer grade C. This answer is correct but is missing some crucial detail. Why is it that nerve cells need the stem cells to regrow? To achieve a B grade you would need to add that nerve cells do not regrow but stem cells can become any cell, given the right biochemical trigger. The hope is that the stem cells can become nerve cells and remake the connection.

Ideas About Science

Understanding the scientific process

As part of your Science assessment, you will need to show that you have an understanding of the scientific process – Ideas about Science.

Science aims to develop explanations for what we observe in the world around us. These explanations must be based on scientific evidence, rather than just opinion. Scientists therefore carry out experiments to test their ideas and to develop theories. The way in which scientific data is collected and analysed is crucial to the scientific process. Scientists are sceptical about claims that cannot be reproduced by others.

You should be aware that there are some questions that science cannot currently answer and some that science cannot address.

Collecting and evaluating data

You should be able to devise a plan that will answer a scientific question or solve a scientific problem. In doing so, you will need to collect and use data from both primary and secondary sources. Primary data is data you collect from your experiments and surveys, or by interviewing people.

While collecting primary data, you will need to show that you can identify risks and work safely. It is important that you work accurately and that when you repeat an experiment, you get similar results.

Secondary data is found by research, often using ICT (the Internet and computer simulations), but do not forget that books, journals, magazines and newspapers can also be excellent sources. You will need to judge the reliability of the source of information and also the quality of any data that may be presented.

Presenting and processing information

You should be able to present your information in an appropriate, scientific manner, using clear English and the correct scientific terminology and conventions. You will often process data by carrying out calculations, drawing a graph or using statistics. This will help to show relationships in the data you have collected.

You should be able to develop an argument and come to a conclusion based on analysis of the data you collect, along with your scientific knowledge and understanding. Bear in mind that it may be important to use both quantitative and qualitative arguments.

You must also evaluate the data you collect and how its quality may limit the conclusions you can draw. Remember that a correlation between a factor that's tested or investigated and an outcome does not necessarily mean that the factor caused the outcome.

Changing ideas and explanations

Many of today's scientific and technological developments have benefits, risks and unintended consequences.

The decisions that scientists make will often raise a combination of ethical, environmental, social and economic questions. Scientific ideas and explanations may change as time passes, and the standards and values of society may also change. It is the job of scientists to discuss and evaluate these changing ideas, and to make or suggest changes that benefit people.

Glossary

A

abundance a measure of how common a species is in an area 26

accuracy how near a reading is to the true value 21

active site part of an enzyme where a substrate can fit neatly into it 23, 28

active transport the movement of chemicals into or out of a cell from areas of low concentration to high concentration, where the cell controls the direction in which chemicals move rather than the difference in concentration 25, 27, 28

adaptation the way in which a species changes over time to become better able to survive in its environment 17, 20, 22

adrenaline a hormone that helps prepare your body for action in the 'fight or flight' response 39

adult stem cells unspecialised body cells that can develop into other, specialised cells that the body needs 8, 9, 29, 33, 34

aerobic respiration respiration that requires oxygen 24, 27–28

algal blooms a thick mat of algae near the surface of water that stops sunlight getting through 55

alleles different versions of a gene on a pair of chromosomes 5–7, 9, 19

amino acids small molecules from which proteins are built 4, 23–24, 32, 34

anaerobic respiration respiration that does not need oxygen 27–28

antagonistic pair two muscles that work in opposite directions 41

antagonistically acting in opposite directions 49

antibiotic therapeutic drug acting to kill bacteria taken into the body 11–12, 16, 19–20

antibody protein normally present in the body, or produced in an immune response, which neutralises an antigen 10, 16

antidepressant a prescribed drug that makes synapses in the brain more sensitive to certain types of transmitter substances 39

anti-diuretic hormone (ADH) hormone which controls re-absorption of water in kidneys (and so water levels in the blood) 15

antigen harmful substance that stimulates the production of antibodies in the body 10–11, 16

antimicrobial substance that acts to kill bacteria 11–12, 16

arteries blood vessels that carry blood from the heart to other parts of the body 13–14, 16

asexual reproduction reproduction (creation of offspring) involving only one parent; offspring are genetically identical to the parent 8, 10

atom the basic 'building block' of an element which cannot be chemically broken down 18, 25–26, 28–29,

average a typical value, used to summarise complicated data sets. For example, the arithmetic mean is produced by finding the total of all the results and then dividing by the number of results to produce a representative figure 43

auxin a plant hormone that affects the rate of growth 30, 34

axon a long projection from a nerve fibre that conducts impulses away from the body of a nerve cell 35, 40

B

bacteria single-celled microorganisms, some of which may invade the body and cause disease 8, 10–12, 16–20 23, 27–28

base one of the three molecules that makes up a single unit of DNA 4, 32, 34

behaviour the way in which an organism reacts to changes in its environment 37–38, 40

beta blockers a prescribed drug that blocks the adrenaline receptors in the synapses and stops the transmission of impulses 39

binary fission simple cell division 10

bioaccumulation the accumulation of a substance, such as a toxic chemical, in various tissues of a living organism 55

biodegradable a material that can be broken down by microorganisms 21

biodiversity the variety in terms of number and range of different life forms in an ecosystem 21–22, 54

biomass the amount of organic material present in an ecosystem, such as a pond; also, the amount of organic material in an organism (usually measured as dry mass). Also, plant material (often waste from other uses) that is used as a source of energy, for example, through burning in an electricity generator 56

blood plasma yellow liquid in blood, in which the blood cells are carried 15

blood pressure the pressure of blood against the walls of the blood vessels 14, 16

body mass index (BMI) a measure of someone's weight in relation to their height, used as a guide to thinness or fatness; values over 30 indicate obesity 43

C

capillaries small blood vessels that join arteries to veins 13, 16

carbon an element that combines with others, such as hydrogen and oxygen, to form many compounds in living organisms 13, 18, 21, 22, 23–28,

carbon cycle the way in which carbon atoms pass between living organisms and their environment 18, 22

carbon dioxide gas whose molecules consist of one carbon and two oxygen atoms, CO_2; product of respiration and combustion; used in photosynthesis; a greenhouse gas 13, 18, 21, 23–24, 27–28

carrier someone who carries a gene but does not themselves have the characteristic 7, 9

cell body the part of a nerve cell that contains the nucleus 35, 37

cell membrane layer around a cell which helps to control substances entering and leaving the cell 24–25, 28

cell sampling removal of a small number of fetal cells, e.g. from the placenta or amniotic fluid, for testing 7

cellulose large polysaccharides made by plants for cell walls 23, 28

central nervous system (CNS) collectively the brain and spinal cord 34, 40

cerebral cortex the outer layer of the brain 38, 40

chlorophyll the green chemical in plants that absorbs light energy 24, 28

chloroplasts structures characteristic of plant cells and the cells of algae where photosynthesis takes place 24, 28

cholesterol chemical needed by the body for the formation of cell membranes, but too much in the blood increases the risk of heart disease 13

chromosomes thread-like structures in the cell nucleus that carry genetic information – each chromosome consists of DNA wound around a core of protein 4–6, 8–9, 31–32, 34

circulatory system a transport system in the body that carries oxygen and food molecules 13, 16, 29

classify put things into groups according to their properties 22

clinical trials scientific testing of drugs, vaccines and medical processes 12

clone organism genetically identical to another 7–9, 30, 34

closed loop process a loop in which the output from one part becomes the input for another with the whole system feeding back on itself 51

coma a very deep sleep 50

combustion process in which substances react with oxygen releasing heat 18, 22

competition result of more than one organism needing the same resource, which may be in short supply 19, 22, 26

compound substance composed of two or more elements which are chemically joined together, for example H_2O 18, 22

conditioned (response) a learned response that occurs when animals link two or more stimuli that are not connected 37

Glossary

conditioned reflex *see* conditioned (response) 37

continuous variation variation in organisms of features that can take any value, for example height 4

contract to get shorter 41

control group in a drugs trial, the group that receives the placebo allowing researchers to assess whether the drug has an effect in the experimental group 12, 16

coronary arteries blood vessels that carry blood away from the heart 13

coronary heart disease (CHD) when arteries that supply the heart muscle gradually become blocked by fatty deposits, preventing the heart from working properly 13

correlation a link between two factors that shows they are related, but one does not necessarily cause the other; a positive correlation shows that as one variable increases, the other also increases; a negative correlation shows that as one variable increases, the other decreases 13, 26

cross-links bonds that link one polymer chain to another 32

cytoplasm a jelly-like substance within a cell where most of the chemical reactions take place 24, 28, 32, 34–35, 40

D

data information, often in the form of numbers obtained from surveys or experiments 11, 13–14, 18–21, 23, 25–26, 30

daughter cells in mitosis a cell splits to form two daughter cells which are identical to each other 29, 31, 34

decompose in chemistry, separation of a chemical compound into simpler compounds 21

decomposer in a food chain, an organism such as a fungus that uses materials from dead or decaying matter 17, 22

decomposition the action of bacteria and fungi to break down previously living material 17, 22

dehydrated the result of a body losing too much water 48

denaturing when an active site is destroyed and the enzyme molecules are broken apart 28

dendrite a short thread of cytoplasm on a neuron, carrying an impulse towards the cell body 35

denitrifying bacteria bacteria vital to the nitrogen cycle, which change nitrates in the soil to nitrogen 18

detritivore in a food chain, an organism such as an earthworm that breaks down dead or decaying matter into smaller particles 21

differentiation the change of an unspecialised body cell into a particular type of cell 8, 29, 33

diffusion the movement of molecules or particles from regions of high concentration to low concentration 25, 28

dislocation A dislocation occurs when the head of one bone is pulled out of its normal position in the socket of another bone at a joint 44

DNA large (polymer) molecule found in the nucleus of all body cells – its sequence determines genetic characteristics, such as eye colour, and gives each one of us a unique genetic code 4, 5, 8–10, 19–20, 22, 24, 28, 31–32, 34

DNA probe a segment of DNA that can link with a particular gene; the probe is usually labelled with a dye or radioactive marker 59

dominant (allele) the allele that is always expressed, irrespective of the other allele in the pair 5–7, 9

double circulation system a circulatory system in which blood travels twice through the heart for each single trip around the body 45

double helix two strands of the DNA molecule face each other in a way that looks like a ladder, these are then twisted around each other to form a double helix – like a spiral staircase 32, 34

E

ecological services services provided by an ecosystem that do not necessarily depend on a single organism, for example, pollination, decay of materials 54

ecosystem the collection of different organisms in an area together with the important non-living factors such as water supply and temperature 53

Ecstasy an illegal drug that affects the working of the chemical transmitter substance in nerve synapses in a similar way to antidepressants 14–16, 39

effector part of the body that responds to a stimulus 15–16, 35–37, 40, 49

efficiency a measure of how effectively an appliance transfers the input energy into useful energy 17

egg female sex cell of an animal 4–6, 8–9, 17, 19, 29, 31, 33–34

embryo an organism in the earliest stages of development which began as a zygote and will become a foetus 6–9, 29, 33–34

embryonic stem cells cells in or from an embryo with the potential to become any other type of cell in the body 8–9, 29, 33–34

environment an organism's surroundings 4–5, 8–9, 15–22, 30, 34–35, 37, 40

enzymes proteins found in living things that speed up or catalyse reactions 4, 9, 23–24, 28, 30, 36

epidemiological studies studies of the patterns of health and illness in the population 13–14

eutrophication the processes that occur when water is enriched with nutrients (from fertilisers) which allow algae to grow and use up all the oxygen; when waterways become too rich with nutrients (from fertilisers) which allows algae to grow wildly and use up all the oxygen 55

evaporate turn from a liquid to a gas, such as when water evaporates to form water vapour 27

evolution change in a species over a long period of time 19–20, 22

excrete to get rid of waste substances from the body 15, 18

extinct a species that no longer survives 17, 19, 21–22

extinction the process or event that causes a species to die out 17, 19, 21–22

F

faeces waste material passed out of the gut of animals 51

family tree diagram chart showing relationships between members of different generations of a family, which can be used to show inheritance of genetic characteristics 6

feral (children) children who have been isolated during their development are said to be 'feral'. Feral means wild or untamed 38

fermentation the conversion of carbohydrates to alcohol and carbon dioxide by yeast or bacteria 27

fermenters equipment that produces the highest rate of fermentation in a microorganism 57

fertilisation the moment when the nucleus of a sperm fuses with nucleus of an egg 5, 7, 31, 33–34

fetus a later–stage embryo of an animal; the body parts are recognisable 7

food web flow chart showing how a number of living things in an environment depend on one another for their food 17, 22

fossil record the information obtained over the years from fossil collections 19–22

functional protein a protein such as an enzyme that speeds up a chemical reaction 4

G

gametes the male and female sex cells (sperm and eggs) 31, 34

gene a section of DNA that codes for a particular characteristic by controlling the production of a particular protein or part of a protein by cells 4–10, 16, 19, 21, 23–24, 28–29, 32–34

gene pool the complete set of alleles in a population; a larger gene pool results in greater genetic variation 19

genetically engineered occurs when the genes in an organism are altered by bringing in genes from other species, using recombinant DNA techniques 57

Glossary

genetic code the information contained in a gene which determines the type of protein produced by cells 24, 32, 34

genetic diversity the differences between individuals (because we all have slight variations in our genes) 21

genetic screening testing large numbers of individuals for a gene, such as a gene for a genetic disorder 7, 9

genetic testing testing an individual for the gene for a genetic disorder 7, 9

genotype an individual's genetic make up, such as whether they are homozygous or heterozygous for a particular gene 4, 7, 9

gland organ that secretes a useful substance 15, 35–36

glucose a simple sugar 23–24, 27–28, 35, 46, 50

H

habitat the physical surroundings of an organism 17, 21–22

haemoglobin a chemical found in red blood cells that carries oxygen 46, 47

heart rate the number of heartbeats every minute 14, 16, 38–39

heterozygous an individual who has two different alleles for an inherited characteristic 5, 9

high blood pressure blood pressure that is consistently abnormally high 14, 16

homeostasis the way the body keeps a constant internal environment 15–16

homozygous an individual who has two identical alleles for an inherited characteristic 5, 9

hormones substances produced by animals and plants that regulate activities; in animals, hormones are produced by and released from endocrine tissue into the blood to act on target organs, and help coordinate the body's response to stimuli 4, 30, 34–36

hypothalamus part of the brain that has several functions, the most important being to link the nervous system to the endocrine system; detects temperature of blood 49

hypothermia a condition caused by the body getting too cold, which can lead to death if untreated 49

I

identification key a way to find a scientific name for an organism by answering yes/no questions 26, 28

immune when a person has resistance to a particular disease 10

immune system a body system which acts as a defence against pathogens, such as viruses and bacteria 10–11, 16, 33

impulse an electrical signal that travels along an axon 33–40

indicator in chemistry, a substance that shows the presence of an acid or an alkali by a change in colour; in biology, a measure of the quality of a natural environment, for example, the number of sensitive species present in an aquatic environment, or the level of pollutants in the air 18, 22

input something put into a system, for example, food and oxygen for animals, mineral nutrients for plants 51

instinctive response behaviour that comes from reflex responses and does not have to be learned 38

insulin hormone produced by the pancreas that promotes the conversion of glucose to glucagon 35–36, 50

interdependence relationship between several organisms that depend on one another 17, 22

isolated separated, as in a strain of bacteria that can be separated from others, or as in an island that is remote from other land masses 20

K

kidney the organ in the body that controls water balance 15–16, 18

L

leukaemia a cancer of the blood-producing cells that causes a massive increase in the number of white blood cells 61

lichen small organism that consists of both a fungus and an alga 18

Life Cycle Assessment an analysis of the environmental impact of a product, including the production of raw materials, its manufacture, packing, transport, use and disposal 21

living indicator a species, the presence of which gives a measure of the quality of an environment; some species, such as the mayfly, are sensitive to pollutants and others are tolerant 18, 22

long-term memory information from our earliest memories onwards, which is stored for a long period of time 39–40

low blood pressure blood pressure that is consistently abnormally low 14

M

mass extinction event the extinction of a large number of species at the same time 21

MDMA 3,4-methylenedioxymethamphetamine, the scientific name for Ecstasy 14–16

mean an average of a set of data 23

meiosis cell division that results in the formation of gametes 31, 34

membrane (of a cell) the layer around a cell which helps to control substances entering and leaving the cell 24–25, 36, 40

membrane cell electrolysis cell that uses a semi-permeable membrane to separate the reactions at the two electrodes, as in the electrolysis of brine 24

memory the storage and retrieval (bringing back or remembering) of information 4, 10–11, 16, 38–40

memory cells white blood cells that form antibodies in response to a particular antigen and retain the ability to make that antibody should re-exposure to the antigen occur later in life 10–11, 16

meristems special regions in a plant where cells are able to divide 29–30, 34

messenger RNA (mRNA) a molecule that copies the base sequence of the DNA and carries it out of the nucleus of the ribosomes 32, 34

metal a group of materials (elements or mixtures of elements) with broadly similar properties, such as being hard and shiny, able to conduct heat and electricity, and able to form thin sheets (malleable) and wires (ductile) 26

microorganism very small organism (living thing) which can only be viewed through a microscope 10–12, 16–18, 21–22

minerals solid metallic or non-metallic substances found naturally in the Earth's crust 25

mitochondria found in the cytoplasm, where respiration takes place 24, 31

mitosis cell division that takes place in normal body cells and produces identical daughter cells 29, 31, 34

molecule two or more atoms held together by strong chemical bonds 4, 9–10, 18, 23–24, 27–28, 31–32, 34, 36, 39–40

monoculture when a single crop is grown 21

motor neuron nerve carrying information from the central nervous system to muscles and glands 35, 37, 40

multicellular consisting of many cells 29, 34–35

multi-store model a type of model used by scientists to help explain how we remember and retrieve information 39–40

mutation a change in the DNA in a cell 5, 12, 19, 22

myelin a fatty sheath that surrounds an axon, it acts as an insulator and makes an impulse travel faster 35, 40

N

nanotechnology technology making use of nanoparticles 60

natural selection process by which characteristics that can be passed on in genes become more common in a population over many generations (which are likely to give the organism an advantage that makes it more likely to survive) 19–20, 22

Glossary

negative feedback information that causes a reversal in a control system, for example when we get too hot our body responds to bring our temperature back to normal through sweating and vasodilation 15, 19, 50

nerve a group of nerve fibres 34–36, 39–40

nervous system sends messages between body cells using neurons; includes the central nervous system (brain and spinal cord) and the peripheral nervous system (network of neurons) 35, 37, 39–40

neuron a nerve cell that carries nerve impulses 35–40

neuron pathway neurones linked to pass nerve impulses 38, 40

neuroscientists scientists who study the nervous system 38

nitrates salts containing the nitrate ion (consisting of one nitrogen atom and three oxygen atoms); may be used as fertilisers, sometimes causing pollution of waterways 18, 22, 24–25, 28

nitrogen cycle the way in which nitrogen and nitrates pass between living organisms and the environment 18, 22

nitrogen-fixing bacteria bacteria vital to the nitrogen cycle, which change nitrogen from the air to nitrates in the soil, needed by plants 18

non-living indicator a non-living measure of the quality of an ecosystem, such as water temperature 18, 22

nucleus a distinct structure in the cytoplasm of cells that contains the genetic material 4, 8–9, 24, 28, 31–35

O

oestrogen female hormone secreted by the ovary and involved in the menstrual cycle 35

open loop process a loop that does not feed back on itself and so is not stable 51

organ a part of the body made up of different tissues that work together to do a particular job 29, 34–37

organic matter material produced by living organisms, typically containing carbon in complex chemicals 51

osmosis the diffusion of water molecules through a partially permeable membrane 26, 28

outlier a measurement that does not follow the trend of other measurements 25

output something produced by a system, often as a waste product, for example, oxygen from a plant in the light 51

overfishing Removing more fish in a season than can be replaced in the same time by reproduction of the fish population 56

oxyhaemoglobin the chemical formed when oxygen combines with haemoglobin 47

P

pacemaker an electronic device used to regulate the beating of the heart 61

pancreas organ that produces hormones insulin and glucagon (from endocrine tissue) and digestive enzymes (from exocrine tissue) 35, 50

partially permeable membrane a cell membrane that lets small molecules pass through but not large ones 25, 28

pathogen harmful organism which invades the body and causes disease 10–11

peripheral nervous system (PNS) network of neurons leading to and from the brain and spinal cord 35, 40

pesticide a chemical designed to kill pests 56

pH a measure of the acidity or alkalinity of a substance 15, 23, 28

phenotype the physical expression of a gene; different genotypes can give the same phenotype 4, 9

phloem plant cells that carry dissolved substances to every part of the plant 29, 34

photosynthesis process carried out by green plants in which sunlight, carbon dioxide and water are used to produce glucose and oxygen 17–18, 22–25, 27–30

phototropism a plant's growth towards or away from the stimulus of light 30, 34

physiotherapist a physiotherapist treats illnesses, usually muscle and bone damage, with exercises and massage rather than drugs 44

phytoplankton microscopic plant life, often forming the basis of aquatic food chains 18, 24

placebo 'dummy' treatment given to some patients in a drug trial, that does not contain the drug being tested 12, 16

planet large sphere of gas or rock orbiting a star 19

plastic a compound produced by polymerisation, capable of being moulded into various shapes or drawn into filaments and used as textile fibres 21

positively phototropic plant shoots that grow towards a light source 30

pre-implantation genetic diagnosis (PGD) genetic testing of embryos created by in vitro fertilisation for a genetic disorder, so that healthy embryos can be transferred into the mother's uterus 7, 9

processing centre a centre of control that acts in response to information, for example the hypothalamus in the brain which responds to changes in body temperature 15–16

products chemicals produced at the end of a chemical reaction 23–25, 29

protein a type of chemical with important functions in living organisms 4, 9–10, 18, 22, 23–25, 28, 32–34

pulse in the body, a beat of the heart that may be felt in an artery close to the skin; in data transmission, a short on-phase 14, 16, 35

pulse rate a measure of the number of times per minute the heart is beating 14, 16

Punnett square a diagram that can be used to work out the probability of outcomes resulting from a genetic cross 6, 9

Q

quadrat a frame, usually square, of wood or metal, that ecologists put on the ground and count the number of plants within it 18, 26, 28

R

rate a measure of speed; the number of times something happens in a set amount of time 12, 14, 23, 25, 28

receptor part of a neuron that detects stimuli and converts them into nerve impulses 15–16, 35–37, 40

receptor molecules found on the membrane of cells, they allow transmitter substances to bind to them 36, 40

recessive (allele) an allele that is only expressed if the other allele in the pair is also recessive; it is hidden if the other allele in the pair is dominant 5–7, 9

recombinant DNA a form of DNA produced by combining genetic material from two or more different sources by means of genetic engineering 58

recovery period the time taken after exercise for breathing and heart rate to return to normal 44

reflex a muscular action that we make without thinking 36–37, 40

reflex arc pathway taken by nerve impulse from receptor, through the nervous system to effector, bringing about a reflex response 37, 40

relay neurons found in the CNS they connect sensory neurons to motor neurons and so co-ordinate the body's response to stimuli 37, 40

reproductive structures parts of a living organism concerned with producing offspring 52

resistance ability of an organism to resist death/disease/ harm, for example resistance may develop in some microorganisms against antimicrobials 11–12, 16, 19

respiration process occurring in living things where oxygen is used to release the energy in foods 15–18, 22, 23–25, 27–28

response the action taken as the result of a stimulus 15, 37–38, 40

Glossary

resting heart rate a person's heart rate when inactive 14

ribosomes small structures in the cytoplasm that make proteins 32

S

salt generically, the dietary additive sodium chloride; in chemistry, an ionic compound formed when an acid neutralises a base 13, 15, 29

saturated fat a component of the diet that, when eaten in excess, can contribute to coronary heart disease and other health problems 13

selective breeding choosing organisms with desired characteristics to breed with one another 19

selective herbicide a herbicide that only kills certain types of plants, leaving others unharmed 58

sensory neuron nerve cell carrying information from receptors to the central nervous system 35

serotonin a transmitter substance found at the synapses in the brain 36

sewage liquid and solid wastes carried in sewers 55

sex cells the male and female gametes (sperm and eggs) 5

sex-determining gene a gene carried on the Y sex chromosome that causes a fetus to develop into a male 6

sexual reproduction reproduction of an organism that involves two parents 5

shell the outer parts of the body, for example, arms and legs, which can tolerate lower temperatures than the core (heart and brain) 49

short-term memory information from our most recent experiences, which is only stored for a short period of time 75

side effects unwanted effects produced by medicines 11–12

signal information that is transmitted by, for example, an electrical current or an electromagnetic wave 36

slash-and-burn a form of agriculture in which an area of forest is cleared by cutting and burning and is then planted, usually for several seasons, before being left to return to forest 54

specialised (cells) a cell that has a particular function 29

species basic category of biological classification, composed of individuals that resemble one another, can breed among themselves, but cannot breed with members of another species 17, 19–21

sperm male sex cell of an animal 5

sprain a sprain occurs when the ligaments around a joint are stretched too much and the joint becomes less stable 44

stable system a system that tends to remain the same over time 52

stem cells unspecialised body cells that can develop into other, specialised cells 8, 33, 61

stimulus (pl stimuli) a change in the environment that causes a response by stimulating receptor nerve cells, e.g. a hot surface 36–37

stores the name given to different types of memory storage in the multi-store model 39

structural protein a protein, such as collagen, whose function is to build tissues 4

substrates the chemicals that enzymes work on 23

superbug harmful microorganism that has become resistant to antimicrobials 12

sustainability measure of whether a resource or process we use now will still be able to be used by future generations 21

sustainable resource or process that will still be available to future generations 21, 54

synovial joint a type of joint that allows a good degree of movement 42

T

temperate the climate of regions between the tropical regions and the polar regions, marked by well-defined cool and warm seasons 54

tendon a flexible, not-elastic tissue connecting bones to muscles 42

theory a creative idea that may explain an observation and that can be tested by experimentation 19–20

therapeutic cloning a procedure in which a nucleus is removed from an egg and is replaced with a nucleus from a body cell in order to produce new cells with identical genes 33

tissue a group of cells that work together and carry out a similar task, such as lung tissue 29

tissue culture (in plants) a method of cloning by taking small pieces of plant tissue from the root or stem and treating it with enzymes to separate the cells, which then grow into separate identical plants 30

toxins poisons or hazardous substances 10, 39

transmitter substance a chemical that passes across a synapse 36

transpiration loss of water from the leaves and stems of plants by evaporation 53

U

urea waste product excreted by the kidneys 15

V

vaccination medical procedure, usually an injection, that provides immunity to a particular disease 11

vaccine weakened microorganisms that are given to a person to produce immunity to a particular disease 11

valve a flap of tissue that allows fluid to pass one way but not the other 45

variation differences between individuals belonging to the same species 4–5, 19

vasoconstriction the narrowing of the lumen (internal space) of blood vessels; in cold conditions, the diameter of small blood vessels near the surface of the body decreases, reducing the flow of blood 48

vasodilation the widening of the lumen (internal space) of blood vessels; in hot conditions, the diameter of small blood vessels near the surface of the body increases, increasing the flow of blood 15, 48

vector an organism that transmits pathogens from host to host – insects are common disease vectors; an animal that carries a pathogen without suffering from it 58

veins blood vessels that carry blood from parts of the body back to the heart 13

virus very small infectious organisms that reproduce within the cells of living organisms and often cause disease; they consist of a protein layer surrounding a strand of nucleic acid 58

W

white blood cell blood cell that defends the body against disease 10

X

xylem cells plant cells that carry water and mineral salts to where they are needed 29, 34

Y

yield the amount of useful material produced by a system 56

Z

zygote the cell formed by the fusion of a male and female gamete at fertilisation 5, 29, 34

Exam tips

The key to successful revision is finding the method that suits you best. There is no right or wrong way to do it.

Before you begin, it is important to plan your revision carefully. If you have allocated enough time in advance, you can walk into the exam with confidence, knowing that you are fully prepared.

Start well before the date of the exam, not the day before!

It is worth preparing a revision timetable and trying to stick to it. Use it during the lead up to the exams and between each exam. Make sure you plan some time off too.

Different people revise in different ways, and you will soon discover what works best for you.

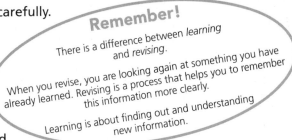

Remember!

There is a difference between *learning* and *revising*.

When you revise, you are looking again at something you have already learned. Revising is a process that helps you to remember this information more clearly.

Learning is about finding out and understanding new information.

Some general points to think about when revising

- Find a quiet and comfortable space at home where you won't be disturbed. You will find you achieve more if the room is ventilated and has plenty of light.

- Take regular breaks. Some evidence suggests that revision is most effective when tackled in 30 to 40 minute slots. If you get bogged down at any point, take a break and go back to it later when you are feeling fresh. Try not to revise when you're feeling tired. If you do feel tired, take a break.

- Use your school notes, textbook and this Revision guide.

- Spend some time working through past papers to familiarise yourself with the exam format.

- Produce your own summaries of each module and then look at the summaries in this Revision guide at the end of each module.

- Draw mind maps covering the key information on each topic or module.

- Review the **Grade booster checklists** on page 143–147.

- Set up revision cards containing condensed versions of your notes.

- Prioritise your revision of topics. You may want to leave more time to revise the topics you find most difficult.

Workbook

The **Workbook** (pages 83–142) allows you to work at your own pace on some typical exam-style questions. These are graded to show the level you are working to (G–E, D–C or B–A*). You will find that the actual GCSE questions are more likely to test knowledge and understanding across topics. However, the aim of the Revision guide and Workbook is to guide you through each topic so that you can identify your areas of strength and weakness.

The Workbook also contains example questions that require longer answers (**Extended response questions**). You will find one question that is similar to these in each section of your written exam papers. The quality of your written communication will be assessed when you answer these questions in the exam, so practise writing longer answers, using sentences. The **Answers** to all the questions in the Workbook can be cut out for flexible practice and can be found on pages 152–160.

Notes

Collins
Workbook

NEW GCSE

Biology

OCR

Twenty First Century Science

Authors: **Eliot Attridge**
John Beeby

Revision Guide +
Exam Practice Workbook

1 Genes help to determine your characteristics and make you who you are.

G–E

a Write down the name of the part of the cell in which genes are found. .... [1 mark]

b Write down the name of the chemical that makes up your genes. .... [1 mark]

2 Write down one example of:

D–C

a a structural protein: .... [1 mark]

b a functional protein: .... [1 mark]

3 The Human Genome Project researched the genes found on human chromosomes.

a Explain **one benefit** of the Human Genome Project.

B–A*

....

.... [2 marks]

b Explain **one ethical consideration** of the Human Genome Project.

....

.... [2 marks]

4 Look at the drawing of Jodie.

pink hair

hair is straight

pale skin colour has been changed by tanning

blue eyes

scar

decayed tooth

pierced ear

dimple

G–E

a Write down **two** characteristics controlled by Jodie's **genes**.

....

....

.... [2 marks]

b Write down **two** characteristics that have been caused by Jodie's **environment**.

....

.... [2 marks]

c Write down one characteristic that is a result of Jodie's genes being affected by her environment. Explain your answer.

.... [2 marks]

5 Write down a characteristic controlled by more than one pair of genes. Describe how this will affect the appearance of the characteristic in the population.

D–C

....

.... [2 marks]

6 Explain the difference between the terms genotype and phenotype for a named characteristic in humans.

B–A*

....

....

.... [3 marks]

7 Here are some statements about identical twins. Put a tick (✓) in the box next to the correct statement.

B–A*

a They have slightly different combination of genes in their cells. ☐

b If separated at birth, they will still look identical as they grow. ☐

c Their appearance can be affected by their environment. ☐

d Their hair colour is a feature always dependent on their genes. ☐ [1 mark]

1 Here are some statements made by students about chromosomes. Some are correct, while some are incorrect. Put ticks (✓) in the boxes next to the **two** correct statements.

a There are 46 chromosomes in human cells. ☐

b Egg cells contain 23 pairs of chromosomes. ☐

c Sex cells contain one chromosome from each pair. ☐

d The genes for a characteristic are in different places on each chromosome of the pair. ☐ **[2 marks]**

G–E

2 Mutations can occur in humans that result in abnormalities in a baby.

a Describe how a chromosome mutation can occur.

....

.... **[2 marks]**

b Write down one example of a chromosome mutation. Genetically, how is this person different from an individual without the chromosome mutation?

....

.... **[2 marks]**

D–C

3 Explain why you are similar to, but different from, your parents.

You are similar because:

....

....

.... **[3 marks]**

You are different because:

....

....

.... **[4 marks]**

B–A*

4 Write down the definition of an allele.

....

.... **[1 mark]**

D–C

5 The following diagrams of chromosomes show the combinations possible for the alleles, with dimples, D, and without dimples, d.

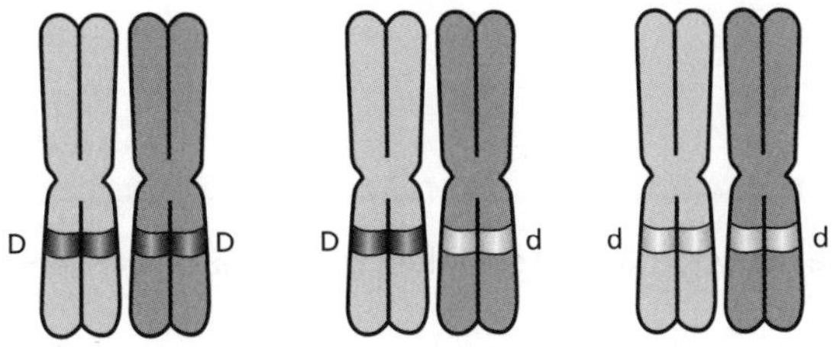

With reference to the diagrams, explain the terms **homozygous** and **heterozygous** for the inheritance of dimples.

....

....

....

.... **[4 marks]**

B–A*

Genetic crosses and sex determination

G–E

1 Use the words provided to complete the sentences. You can use the words once, or not at all.

alleles	**bases**	**chromosomes**	**different positions**
DNA	**each chromosome**	**identical positions**	**one chromosome**

Genes for a particular trait are located at ....

on .... of the chromosome pair. Different versions of

genes are called .... **[3 marks]**

D–C

2 People are either able to roll their tongue into a U-shape, or unable to roll their tongue. The diagram shows a pair of chromosomes.

Genotype TT: can roll tongue

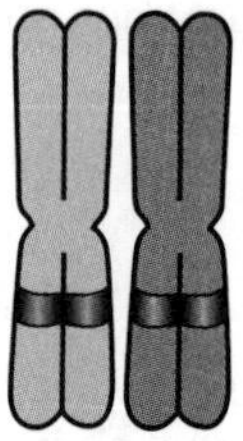

a Is the allele for tongue rolling dominant or recessive? How do you know?

.... **[2 marks]**

b Write down the other possible genotypes related to tongue rolling.

.... **[2 marks]**

c Write down the phenotypes for these genotypes.

.... **[2 marks]**

B–A*

3 A man and woman who can both roll their tongues have children. Use the information in the genotype diagram above to answer the questions.

a Write down the possible genotypes of the couple. .... **[2 marks]**

b The couple have children. Their first child cannot roll his tongue; the second one can.

i What does this tell you about the genotypes of the couple? Explain your answer.

....

.... **[3 marks]**

ii Show the genetic cross involved. Use a Punnett square to illustrate your answer.

[3 marks]

D–C

4 Explain why some embryos develop into male babies.

....

....

.... **[3 marks]**

B–A*

5 Explain why males are more likely to have a sex-linked disease such as haemophilia or colour blindness.

....

....

....

.... **[5 marks]**

1 Some diseases are caused by defective genes. Use the words provided to complete the sentences.

concentrating   cystic fibrosis   digesting food

Huntingdon's disease   recessive   dominant

a For some genetic disorders a single allele is sufficient to cause the disorder. These are

called .... disorders. One example of this is ...., where symptoms

include tremors, memory loss and difficulty in ....   [3 marks]

b In other types of genetic disorder, two alleles are needed for the disease to occur. These are

called .... disorders. One example is ...., where symptoms include

difficulty in breathing and difficulty in ....   [3 marks]

2 About 1 in every 10 000 babies born will have inherited the allele for Huntington's disease.

In the UK, around 700 000 babies are born every year. How many of these would you expect
to have inherited the allele for Huntington's disease? Show your working.

....

....

....   [2 marks]

3 Genetic testing is carried out on adults, children and unborn fetuses.

a Write down two reasons for carrying out genetic testing on **children** or **adults**.

....

....   [2 marks]

b Write down two genetic testing techniques used during pregnancy.

....

....   [2 marks]

4 Genetic screening programmes have advantages, but they also have ethical implications.

a Explain the implications of genetic screening programmes carried out:

i on week-old babies in the heal prick blood test:

....

....   [2 marks]

ii by employers and insurance companies:

....

....   [2 marks]

b Explain the advantages of testing embryos (by pre-implantation genetic diagnosis) before
implanting the embryos following in vitro fertilisation.

....

....

....   [2 marks]

B1 You and your genes   87

1 There are advantages and disadvantages to asexual reproduction.

a Explain **one** advantage. .... **[1 mark]**

b Explain **one** disadvantage. ....

.... **[2 marks]**

2 Describe the technique used to clone Dolly the sheep artificially.
Annotate the diagram to illustrate your answer.

Dolly's genetic mother

cloned sheep born after normal pregnancy

Dolly

[4 marks]

3 Use the words provided to complete the sentences on stem cells.

adult   any   certain   diseases   embryo   embryonic   unspecialised

.... stem cells are found in the five-day old .... These cells can

produce .... cell type in the human body because they are ....

Stem cells remain in our bodies as grown-ups. They too can be used to treat .... as

they can develop into .... cell types. **[6 marks]**

4 Several types of stem cell are found in humans.

a Explain the differences between embryonic and adult stem cells.

....

....

.... **[4 marks]**

b Read the following sentence: 'Scientists in Japan have reprogrammed human skin cells so that they behave like embryonic stem cells and have the potential to become any tissue in the body.'

Explain why these reprogrammed skin cells would be less controversial than the use of embryonic stem cells in the treatment of disease.

....

.... **[3 marks]**

5 Explain the potential of stem cells in medicine.

....

....

....

.... **[4 marks]**

A couple visit their doctor. Their first child was born with a single gene disorder called phenylketonuria (PKU). The phenylketonuria allele is recessive. The couple are both healthy, but the disorder has been seen before in both families.

The doctor first explains to the family how, as healthy individuals, they could have produced a child with phenylketonuria. She draws a Punnett square to help with this explanation.

The couple see the doctor's diagram and think because they've had one child with PKU, there will be less chance, next time, of having a child with the disorder. The doctor explains why this isn't true.

Use your knowledge and understanding of genetics to write out the explanations that the doctor gives to the couple.

🖉 *The quality of written communication will be assessed in your answer to this question.*

[6 marks]

1 Here are some statements about microorganisms and disease. Some statements are correct, while some are incorrect. Put ticks (✓) in the boxes next to the **two** correct statements.

Bacteria make us feel ill by releasing toxins. ☐

Microorganisms that cause disease are called pathogens. ☐

The only microorganisms that cause disease are bacteria and viruses. ☐

Diseases caused by viruses include colds and TB. ☐ [2 marks]

2 A person has an accident and 10 cells of the flesh-eating bacterium *Streptococcus pyogenes* enter the wound. At human body temperature, the flesh-eating bacterium can divide into two every 10 minutes. For every million bacteria, 1 cm² of flesh is eaten away. What area of flesh would be eaten 3 hours after infection?

....

....

.... [2 marks]

3 The diagram below shows the growth curve of a bacterium.

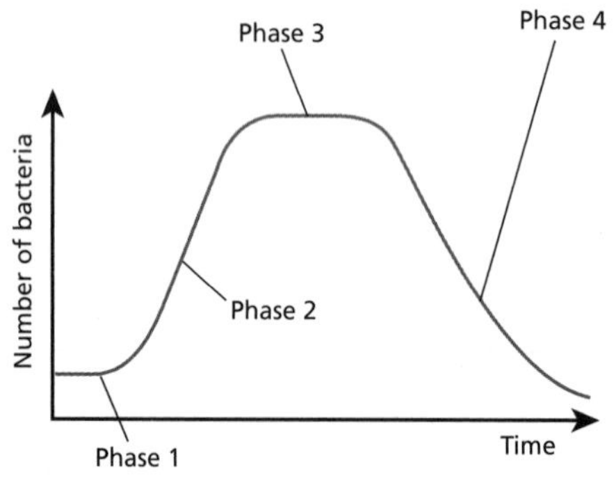

a On the diagram, label the four phases of bacterial growth. [1 mark]

b Describe what is happening in each phase.

Phase 1: ....

Phase 2: ....

Phase 3: ....

Phase 4: .... [4 marks]

4 Here are some statements about ways in which the body has external barriers to microorganisms. Choose from the words provided to complete the sentences.

acid   alkali   intestines   oil

saliva   skin   stomach   tears

Our .... is able to heal itself, and produces ...., which can stop

bacteria from reproducing. Chemicals in our .... and .... also have

antibacterial properties. Our .... produces ...., which destroys many

bacteria in our digestive system. [6 marks]

1 Here are some statements made by students about vaccines. Put a tick (✓) in the box next to the **one** correct statement.

Vaccines are always made from dead microorganisms. ☐

The microorganisms in vaccines have antibodies on their surface. ☐

Memory cells are left in the blood after vaccination. ☐

An infecting microorganism is destroyed very quickly by antigens in the blood. ☐

[1 mark]

G–E

2 Courses of vaccinations are given to babies and children.

   a Write down what a vaccination against a microorganism does.

.... **[1 mark]**

   b Explain why a course of vaccinations is given to babies and children.

.... **[1 mark]**

D–C

   c Explain why a single flu vaccine would not give permanent protection against the disease, and new ones have to be developed.

....

....

.... **[3 marks]**

3 After a science lesson about vaccinations, some friends are discussing the use of vaccines to eradicate disease. Here are some quotes:

William: 'Using vaccines, we've already eradicated one disease from the world.'

Xavier: 'We could do this for others if we vaccinated the whole population.'

Yvonne: 'But vaccines are very often unsafe.'

Zak: 'And we can't vaccinate against all diseases.'

Which statements are correct? Use these ideas to discuss whether vaccinations could eradicate infectious diseases globally.

B–A*

....

....

....

....

....

.... **[5 marks]**

4 Some people get unwanted reactions when they are given a vaccination or a prescription drug.

   a What is this type of unwanted reaction called?

.... **[1 mark]**

G–E

   b Why is this type of reaction more severe in some people than others?

.... **[1 mark]**

5 Write down two effects of antimicrobials that are effective on microorganisms.

....

B–A*

.... **[2 marks]**

1 a Some bacteria develop resistance to antibiotics. Explain how this resistance develops.

....

....

.... **[5 marks]**

b Write down three recommendations for the use of antibiotics.

....

.... **[3 marks]**

2 The resistance that some bacteria show to antibiotics was investigated in Belgium between 1985 and 2008.

– The graph below left shows the resistance of the bacterium that causes pneumonia to four antibiotics.

– The graph below right shows the total antibiotic use between 1997 and 2007, in packages per 100 000 inhabitants per day.

(Source: National Reference Centre S. *pneumoniae*, University *Leuven*)

(Source: Goossens *et al.* (2008). Achievements of the Belgian Antibiotic Policy Coordination Committee. *Eurosurveillance* **13**, 10–13)

a Describe the trends shown in resistance of the bacterium to the antibiotics in the left-hand graph.

....

....

.... **[5 marks]**

b What can you conclude from the graph of total antibiotic use?

.... **[1 mark]**

c What can you conclude from the two graphs together?

....

.... **[2 marks]**

3 Describe **two** types of clinical trials.

....

.... **[2 marks]**

4 Three types of clinical trial used are open-label, blind study and double-blind study. Explain why the double-blind test is considered to generate the most reliable results.

....

....

.... **[3 marks]**

1 Use the words provided to complete the sentences about the heart and circulatory system. You can use the words once, or not at all.

blood   double pump   half   heart   nutrients   oxygen

quarter   single pump   veins   wastes   lungs   blood vessels

a The blood carries .... and .... to the body's cells.

It removes .... from the body's cells. **[3 marks]** G–E

b The heart acts as a .... As one .... of the heart is carrying blood

to the body, the other is carrying blood to the .... **[3 marks]**

c The circulatory system is made up of the ...., ... and

.... **[3 marks]**

2 Complete the table to explain the structure and function of arteries, capillaries and veins.

	Function	How their structure is related to function
Arteries		
Capillaries		
Veins		

D–C

[6 marks]

3 Heart disease kills more people in the UK than any other disease. Explain what coronary heart disease is.

....

.... **[2 marks]**

G–E

4 Trends in heart disease are studied across the world. The following graph shows male deaths from CHD for different age groups, from 1968 to 2008.

(Source: The British Heart Foundation)

B–A*

a Describe the trend in heart disease in men, from 1968 to 2008.

....

.... **[2 marks]**

b In which group has the change been greatest? .... **[1 mark]**

c Suggest three factors that could have led to this change.

....

.... **[3 marks]**

d Trends in data suggest that the more deprived the socio-economic group, the greater the risk of death from CHD. How might scientists look for evidence of factors that may be involved?

.... **[1 mark]**

1 Write down **two** possible consequences of leaving high blood pressure untreated.

....

.... **[2 marks]**

2 Write down **three** lifestyle factors that can lead to heart disease.

....

.... **[3 marks]**

3 Here are some statements about the measurements scientists make of our cardiovascular fitness.

Some statements are correct, while some are incorrect. Put ticks (✓) in the boxes next to the **two** correct statements.

Pulse rate is measured in beats per minute. ☐

Blood pressure is measured using a sphygmomanometer. ☐

The resting heart rate in teenagers is 50–70. ☐

Blood pressure measurements indicate the pressure in our veins. ☐

The fitter we are, the faster our pulse rate during exercise. ☐ **[2 marks]**

4 One epidemiological study carried out in the USA looked at the effects of passive smoking (being exposed to and inhaling someone else's cigarette smoke) on groups of female nurses. The results are shown below.

Type of coronary heart disease (CHD)	Number of cases		
	Never exposed to cigarette smoke	Occasional exposure to cigarette smoke	Regular exposure to cigarette smoke
Non-fatal heart attack	14	63	50
Fatal heart attack	3	11	11

(Source: Kawachi I., Colditz G.A., Speizer F.E, Manson JoAnn, Stampfer M.J. Willett W.C, and Hennekens C.H. (1997). A Prospective Study of Passive Smoking and Coronary Heart Disease. *Circulation* 95 2374-2379)

a What was the total number of CHD cases in the study:

 i in nurses not exposed to cigarette smoke? .... **[1 mark]**

 ii in nurses exposed to any amount of cigarette smoke? .... **[1 mark]**

b How many times more likely to get CHD were the nurses who were exposed to smoke, compared with those exposed to no smoke?

.... **[1 mark]**

c What conclusions can be drawn from the study about the effect of passive smoking on non-fatal heart attacks?

....

....

.... **[2 marks]**

d Why are epidemiological studies of this type difficult to carry out?

....

....

....

.... **[3 marks]**

1 Write down a definition of **homeostasis**.

.... [1 mark]

2 A person develops hypothermia. Which factor in the body has been allowed to change? Put a tick (✓) in the box next to the correct answer.

Blood pressure ☐   Water ☐

Blood sugar ☐   Temperature ☐   Salt ☐   [1 mark]

3 When running, Jane's body produces more carbon dioxide. Complete the flow chart below to show the control systems involved in homeostasis.

[_____] ➡ [_____] ➡ [_____]   [3 marks]

4 a Atiq's body temperature rises during strenuous exercise. On the diagram below, complete the labels to show how the changes in his body return his temperature to normal.

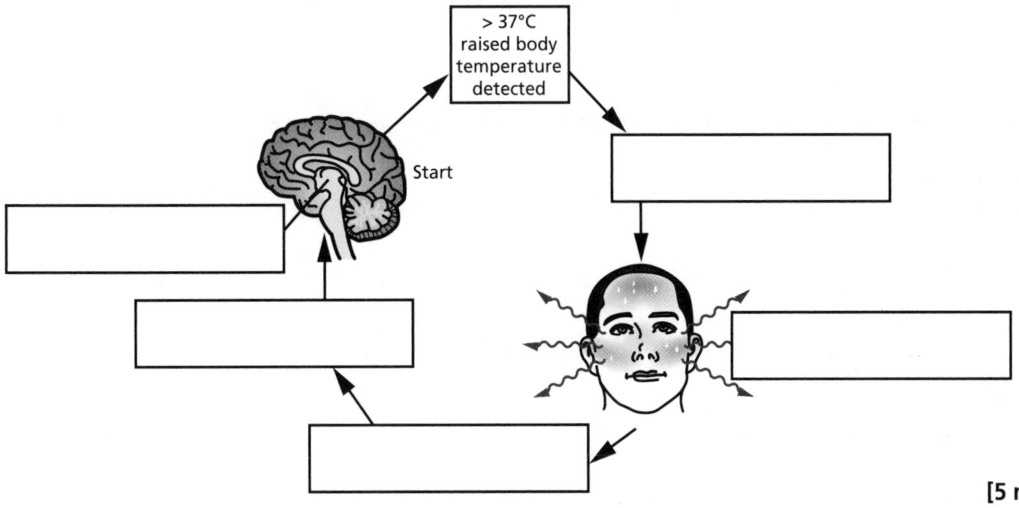

> 37°C raised body temperature detected

Start

[5 marks]

b What happens when Atiq's body temperature returns to normal?

....

.... [2 marks]

c Name the control mechanism that reverses the changes that have put Atiq's body system off balance?

.... [1 mark]

5 Explain what happens to our blood plasma and urine if we:

a drink a lot of water: ....

.... [3 marks]

b eat a lot of salt: ....

.... [3 marks]

6 a Anti-diuretic hormone (ADH) helps us to control the water balance of our bodies. How does ADH work?

.... [2 marks]

b Explain the effect that alcohol has on the water balance of the body.

....

.... [3 marks]

Gary sees this article in a magazine.

In Europe prior to 1940, the numbers of deaths from many diseases, including diphtheria, were negligible. But after mass vaccination programmes against diseases began, wave after wave of epidemics followed in fully vaccinated people, with many deaths. Vaccinations are harmful!

Gary does not think that this is correct. He does some research and finds this table in the Department of Health's Vaccination 'Green Book'.

Disease	Total deaths in 2003	Deaths in the UK before the vaccination was introduced (year of introduction)
Diphtheria	0	2133 (1939)
Hib	14	24 (1991)
Measles	0	99 (1967)
Meningococcal C	19	101 (1998)
Polio	0	241 (1955)
Whooping cough	2	92 (1956)

(Source: The Department of Health (2006). Immunisation against infectious disease. ('The Green Book'). The Department of Health.)

Use the data to discuss the claims made in the book.

✎ *The quality of written communication will be assessed in your answer to this question.*

[6 marks]

1 Species of plants and animals are adapted to their environment.

a Write down a definition of a species.

.... [1 mark]

b A cactus is adapted to live in hot, dry conditions. Write down three adaptations and explain how they enable the cactus to live there.

i ....

ii ....

iii .... [3 marks]

G–E

2 The diagram below shows a typical food chain of a moorland.

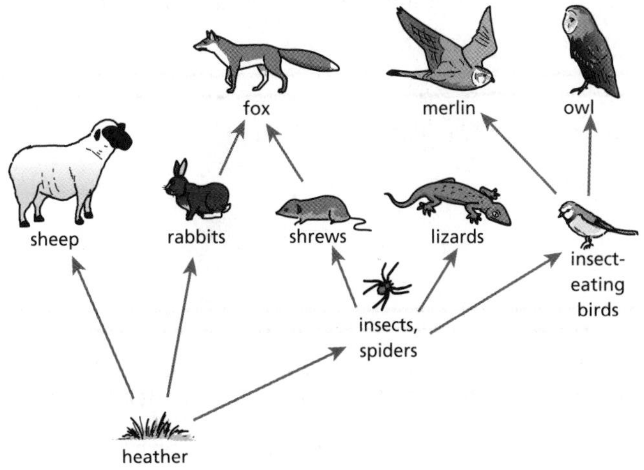

A high concentration of insecticide sprayed onto farmland near an area of moorland ends up on the moorland and kills most of the insects and spiders. Describe what will happen to:

i the foxes ....

.... [2 marks]

ii the insect eating birds ....

.... [2 marks]

iii the lizards ....

.... [3 marks]

iv the owls ....

.... [2 marks]

B–A*

3 Write down one reason why species become extinct.

.... [1 mark]

G–E

4 a In a grassland food web, grasshoppers transfer 111 MJ of energy into their bodies during the summer. Of that energy, 94 MJ is lost during respiration and in faeces. What is the efficiency of energy transfer?

.... [1 mark]

D–C

b When the grasshoppers die, their remains store 16 MJ of energy. Explain what happens to this energy on their death.

.... [1 mark]

B–A*

1 Carbon is recycled through the environment in the carbon cycle. Choose from the words provided to complete the sentences.

air   bacteria   carbon dioxide   combustion   decomposition

fixed   fungi   photosynthesis   respiration   soil

a Carbon enters the carbon cycle as .... from the .... Plants use this

to produce food by the process of .... Carbon is said to be .... by

this process.   **[4 marks]**

b When obtaining energy from food, organisms produce carbon dioxide by the process

of .... As organisms die and are broken down, this gas is also returned to the

air by a process of .... This involves microorganisms such as ....

and .... that live in the ....   **[5 marks]**

G–E

---

2 Explain the processes of nitrogen-fixation and denitrification.

B–A*

....

....

....   **[2 marks]**

---

3 Here are some statements about living and non-living indicators to monitor environmental change. Some are correct, while some are incorrect. Put ticks (✓) in the boxes next to the two correct statements.

Lichens are biological indicators that give information on air temperature.   ☐

Mayfly larvae are good biological indicators because they can only live in clean water.   ☐

Nitrates in the water and air are non-biological indicators that are monitored.   ☐

Monitoring carbon dioxide gives information on climate change.   ☐   **[2 marks]**

D–C

---

4 The following graph shows the carbon dioxide concentration in the air measured at the monitoring station of Mauna Loa, in Hawaii.

Weekly average CO_2 concentrations derived from continuous data at Mauna Loa, Hawaii

CO_2 concentration (ppmv)

Date
(year, month and day, e.g. 760610 indicates June 10th, 1976)

(Source: Keeling, C.D. and T.P. Whorf. (2004) Atmospheric CO_2 concentrations derived from flask air samples at sites in the SIO network. *In Trends: A Compendium of Data on Global Change.* Carbon Dioxide Information Analysis Center, Oak Ridge, USA.)

a Describe how levels in carbon dioxide in the air change from 1958 to 2001. Suggest why these changes have occurred.

....

....

....

....   **[4 marks]**

b Scientists also measure levels of carbon dioxide from other sources. Write down two of these sources.

....

....   **[2 marks]**

B–A*

1 Choose from the words provided to complete the sentences:

complex evolution genetics million simple thousand variation

Life began around 3500 .... years ago with .... forms of life.

More .... forms of life developed later. The process by which new species emerge

over a very long period of time is called .... The changes involved in this process

have come about by .... between individuals. **[5 marks]**

G–E

2 Here are some statements about fossils. Put a tick (✓) in the box next to the **one** correct statement.

It is not usually possible to find out the age of fossils. ☐

The fossils we find are mostly of animals that are alive today; few are extinct. ☐

Fossils are the remains of dead plants and animals. ☐

Fossils of eggs that have been found are where the eggs have been turned into rock. ☐ **[1 mark]**

---

3 Explain what a mutation is, and how mutations can lead to changes in a population of organisms over a period of time.

....

....

.... **[5 marks]**

D–C

4 Explain how changes in the gene pool lead to evolution.

....

....

....

.... **[5 marks]**

B–A*

---

5 An animal breeder is trying to develop a new breed of cattle that produces large volumes of lower fat milk. Explain how the cattle breeder would use selective breeding to achieve this.

....

....

.... **[4 marks]**

D–C

---

6 Scientists in Germany investigated the effectiveness of the poisons warfarin and bromadiolone on rats. Some of their results are shown in the table opposite.

a Explain how the data are evidence for natural selection.

Town	Not resistant to either poison (%)	Resistant to warfarin (%)	Resistant to both poisons (%)
Dorsten	44	56	0
Drensteinfurt	90	5	5
Ludwigshafen	100	0	0
Olfen	21	21	58
Stadtlohn	5	8	87

(Source: Kohn M.H, Pelz H-J, and Wayne R.K (2000). Natural selection mapping of the warfarin-resistance gene. *Proc. Natl. Acad. Sci. USA.* **97** 7911–7915)

B–A*

....

....

.... **[4 marks]**

b Suggest which town had used the rat poison the most prior to the study. Explain your answer.

....

.... **[2 marks]**

1 The following statements are the stages in the development of a new species. They are not in the correct order. Put numbers in the boxes to place the sequence into order.

a The environments are different, so natural selection operates differently on the populations. ☐

b Two populations of a species become isolated by a physical barrier. ☐

c Individuals from the different populations become unable to reproduce with each other. ☐

d Different alleles become more frequent in different populations. ☐

e A new species has been produced. ☐ [5 marks]

D–C

2 Explain how Darwin's observations of the animal and plant life on the Galapagos Islands, volcanic islands off the coast of South America, led to his 'theory of evolution by natural selection'.

B–A*

....

....

....

.... [4 marks]

3 Choose from the words provided to complete the sentences about classifying organisms.

ancestor   appearance   classified   DNA   evolved   groups

Organisms are .... by putting them into .... Previously, many

scientists grouped organisms mainly on their ...., but now they look at the

organisms' .... Using this type of grouping helps us to understand how

organisms .... over thousands or millions of years. [5 marks]

G–E

4 Explain how evidence from the fossil record and DNA provides evidence for evolution.

D–C

....

....

....

.... [3 marks]

5 A French scientist called Lamarck thought that animals changed during their lifetime, and these characteristics would be passed on to their young. For example, as a giraffe stretched to reach leaves on a tree, its neck would get longer. The giraffe would pass on its longer neck to its offspring.

Explain why Darwin's description of the origin of new species is a better explanation.

B–A*

....

....

....

....

.... [5 marks]

1 Here are some statements about biodiversity. Put a tick (✓) in the box next to the **one** correct statement.

A habitat with a narrow range of species of organisms has a high biodiversity. ☐

Biodiversity includes a wide range of genetic variation. ☐

A rainforest has very high biodiversity. ☐

A plantation of tropical oil palm trees has very high biodiversity. ☐

The term biodiversity refers to the range of different groups of organisms. ☐

G–E

[1 mark]

2 Put a tick (✓) in the box next to the **two** correct examples of sustainability.

Meeting our current needs without depriving future generations ☐

Reducing the planet's biodiversity ☐

Recycling and re-using items instead of throwing them away ☐

Removing rainforest for housing ☐

Increasing the amount of land used to grow crops ☐

G–E

[1 mark]

3 Explain why the removal of one species will have a big impact on the ecosystem.

....

....

....

....

D–C

[4 marks]

4 Intensive crop production has involved large-scale planting of one type of crop. This practice is called monoculture.

a Explain why monoculture has negative effects on biodiversity and is non-sustainable.

....

....

....

B–A*

[3 marks]

b Describe the methods scientists, working with farmers, have used to improve biodiversity.

....

....

....

[3 marks]

5 Write down three ways in which sustainability can be improved in the manufacture of a product.

....

....

....

D–C

[3 marks]

6 Explain why it is preferable to decrease the amount of packaging, even if it is biodegradable.

....

....

....

B–A*

[2 marks]

Competition between plants is important to crop growers.

In one study, scientists investigated the growth of banana plants grown at different densities in three different plots. Each plot had an area of three hectares. The following results were obtained

Plot	Number of plants grown in plot	Average plant height in m	Time to reach harvest in months	Average mass of a bunch of bananas in kg	Total yield of bananas in tonnes per hectare
1	5 000	3.5	15	15.3	23
2	10 000	4.2	18	14.3	40
3	15 000	4.3	20	13.3	52

Analyse the data and suggest reasons for the differences between the plants grown in the three plots.

✎ *The quality of written communication will be assessed in your answer to this question.*

[6 marks]

1 Read the statements below about the chemical reactions of living things. Some statements are correct, while some are incorrect. Put ticks (✓) in the boxes next to the **two** correct statements.

a Only animals carry out respiration; plants photosynthesise instead. ☐

b The end producers of photosynthesis are carbon dioxide and water. ☐

c Respiration is the process by which organisms release energy from food. ☐

d During photosynthesis, light energy is converted to chemical energy. ☐

[2 marks]

2 Write down the definition of an enzyme.

.... [1 mark]

3 Here are some statements about enzymes. Use the words provided to complete the sentences.

acid   active site   amino acids   genes   lock and key

product   proteins   optimum   substrate   sugars

Enzymes are .... They are made up of long chains

of .... joined together. Enzymes are assembled according

the instructions in our ....

A molecule that is broken down by an enzyme is called the ....

The part of the enzyme that links with the molecule to be broken down is called

the ....

This model of enzyme action is called the .... model.   [6 marks]

4 The graphs show the effect of pH on the activity of two proteases (enzymes that break down proteins).

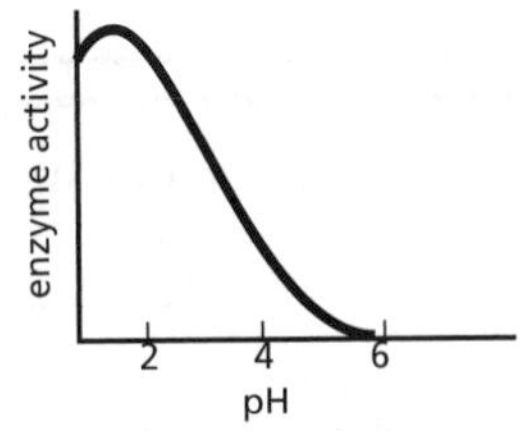

stomach protease

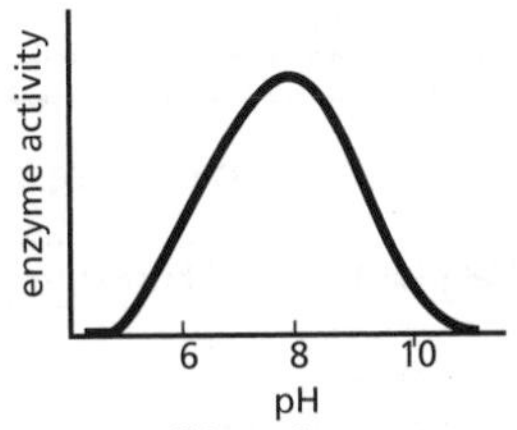

small intestine protease

a What are the optimum pHs of stomach and small intestine proteases?

.... [2 marks]

b Explain why enzyme activity is sensitive to pH.

....

....

.... [3 marks]

5 Olivia finds a graph on the Internet showing how pH affects the activity of a protease called papain, from the papaya plant.

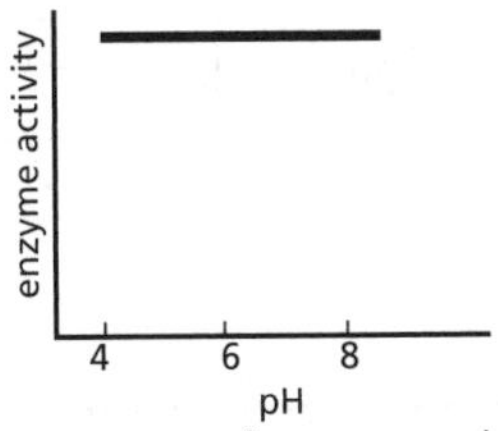

a Olivia says that the graph shows that papain is unaffected by pH. Is Olivia correct? Explain your answer.

....

....

....

.... [2 marks]

protease from papaya plant

b In terms of enzyme structure, suggest what these results show.

.... [1 mark]

B4 The processes of life

1 Fill in the spaces below to complete the word equation for photosynthesis.

$$\boxed{}$$

$$\boxed{} \; + \; \boxed{} \quad \longrightarrow \quad \boxed{} \; + \; \boxed{}$$

[2 marks]

2 Here are some statements about photosynthesis. Use the words provided to complete the sentences below.

amino acids	cellulose	chlorophyll	chloroplasts	glucose
proteins	respiration	starch	sugar	the Sun

The energy for photosynthesis comes from .... The energy is

absorbed by the green pigment called .... This is found in

structures in the cells called ....

The main product of photosynthesis is a .... called

.... Some is used for .... ,

to release energy. Some is converted into large molecules the plant needs, such

as .... , ... and

.... [7 marks]

3 Write down the symbol equation for photosynthesis.

$$\boxed{}$$

$$\boxed{} \; + \; \boxed{} \quad \longrightarrow \quad \boxed{} \; + \; \boxed{}$$

[2 marks]

4 Complete the table below to describe the function of parts of the plant cell.

Part of the plant cell	Function
a Cell membrane	
b Cell wall	
c Cytoplasm	
d Mitochondrion	
e Nucleus	

[5 marks]

5 Complete the table below comparing cell structure and function in microorganisms.

	Bacteria	Yeast
Outer layer of cell		
Genetic code		
Respiration		

[3 marks]

The processes of life

1 Write down a definition of diffusion.

....

.... **[2 marks]**

G–E

2 Here are some statements about the movement of gases in and out of plants. Use the words provided to complete the sentences:

active   carbon dioxide   decrease   increase   into   diffusion

nitrogen   osmosis   out of   oxygen   passive

.... gas, required by the plant for photosynthesis, moves

.... the plant by the process of ....

By the same process, ... gas, produced by photosynthesis,

moves .... the plant.

The process is .... , so if the temperature is increased, the rate at

which the process will take place will .... **[7 marks]**

D–C

3 In the school lab, Adam cut and weighed two potato chips. He placed one in distilled water, and one in concentrated sucrose solution.

a After 30 minutes, Adam weighed the potato chips again. How did the mass of the chips change? Explain your answer.

....

....

.... **[5 marks]**

b Adam wanted to find the concentration of dissolved chemicals in the potato chips. Describe how he could extend his experiment to do this.

....

....

.... **[5 marks]**

4 Explain why nitrates are taken up into plant roots by active transport.

....

.... **[3 marks]**

B–A*

5 The graph opposite shows the effect of light intensity on photosynthesis in a single-celled plant.

a Describe and explain the effect of light intensity on the plant.

....

....

....

.... **[4 marks]**

b The investigation was also carried out in a high concentration of carbon dioxide.

i Sketch a graph of what you would expect on the axes above.

ii Explain its shape.

....

.... **[2 marks]**

D–C

1 Ruby is carrying out a survey of the trees in a wood. Describe and explain how she should use a key to identify the trees from their leaves.

....

....

....

.... [4 marks]

2 Ecologists sample the plants in an area to find out their abundance.

a Quadrats are used in several sizes. For instance:

5 cm × 5 cm   20 cm × 20 cm   0.5 m × 0.5 m   0.5 km × 0.5 km

Which size quadrat would be most suitable for sampling:

i green algae growing on a tree trunk? .... [1 mark]

ii trees growing in a wood? .... [1 mark]

iii daisies growing on the school playing fields? .... [1 mark]

b The diagram shows the results of a survey to estimate the distribution of dandelions in an area. The quadrats are 0.5 m × 0.5 m.

Estimate the distribution of dandelions as plants per m².

.... [2 marks]

c Ecologists sometimes place quadrats on transect lines to carry out their surveys. Write down **one** situation in which they would do this.

.... [2 marks]

3 In an investigation on the number of species of plant identified growing around the base of an oak tree, the following results were obtained.

Distance from oak tree (m)	0	5	10	15	20	25	30
Number of plant species	0	2	8	14	22	33	34

a A student produced a hypothesis that this could be the result of the shade from the oak tree.

Explain how light intensity might affect plant growth.

....

....

....

.... [4 marks]

b On measuring the light intensity at different distances from the tree, the increase in light intensity correlated with the increase in number of plant species present.

What conclusions can be drawn about the hypothesis?

....

....

....

.... [4 marks]

1 Here are some statements about **aerobic respiration**. Some statements are correct, while some are incorrect. Put a tick (✓) in the boxes next to the **two** correct statements.

a Glucose is a product of respiration in plants. ☐

b Some plants carry this out in the absence of oxygen. ☐

c It provides the energy essential for our cells to live. ☐

d One product is carbon dioxide. ☐

e Body cells use starch and protein for aerobic respiration. ☐

G–E

[2 marks]

2 Fill in the spaces below to complete the word equation for respiration.

☐ + ☐ → ☐ + ☐ + ☐

D–C

[2 marks]

3 Aerobic respiration is an essential process in cells.

a Write down the symbol equation for aerobic respiration.

.... [2 marks]

b Why does this equation not fully explain the process?

....

.... [2 marks]

B–A*

4 Write down **three** situations in living organisms in which anaerobic respiration takes place.

....

....

.... [3 marks]

G–E

5 Chemical products of anaerobic respiration in microorganisms are important in certain food products.

a Write down **three** chemical products produced by anaerobic respiration in microorganisms.

....

.... [3 marks]

b Write down **one** type of microorganism that respires anaerobically, and the food product it is used to make.

....

.... [2 marks]

D–C

6 Write down **three** differences between aerobic respiration and anaerobic respiration. In your answer, refer to:

– the use of oxygen

– the products of the reaction

– the amount of energy released.

....

....

....

.... [3 marks]

B–A*

B4 The processes of life 107

An enzyme called invertase is often used when brewing beer. It is added to the mixture of brewing ingredients at the beginning of the process.

Invertase breaks down a sugar called sucrose in the mixture, into the sugars glucose and fructose.

Explain:

- how the enzyme invertase works
- why invertase will not break down the sugar maltose, which is also found in the mixture of brewing ingredients
- why pH 4.5 is the optimum for brewing beer.

Use diagrams to illustrate your answer.

✏ *The quality of written communication will be assessed in your answer to this question.*

D–C

[6 marks]

1 Here are some statements about cell specialisation in animals. Use the words provided to complete the sentences.

brain   multicellular   nerve cells   nervous tissue   organs

single-celled   specialised   tissue   tissues   unspecialised

a In organisms that are ...., cells are

.... to do different jobs.   **[1 mark]**

b Cells of the same type are grouped into a ....

One example of this is ...., which are grouped together

to form ....   **[1 mark]**

c Different .... are grouped together, and work together,

in structures called .... One example of this is

the ....   **[1 mark]**

2 Explain how cells in a human embryo become specialised to do a particular job.

....

....

....

....   **[4 marks]**

3 Here are the names of some plant tissues and organs:

flower   **phloem**   **root**   **stem**   **xylem**

Put these in the correct columns in the table below.

Plant tissue	Plant organ

[5 marks]

4 Here are some statements about cell specialisation and growth in plants. Some statements are correct, while some are incorrect.

Put ticks (✓) in the boxes next to the **two** correct statements.

a Plants keep growing throughout their lives. ☐

b Plant cell division occurs in meristems. ☐

c Plant cell division occurs throughout the shoot and root. ☐

d Meristems produce growth of plant height only. ☐   **[2 marks]**

5 Write down **four** areas where meristems are found in plants. For each of these areas, describe how the meristem produces growth.

....

....

....

....

....   **[4 marks]**

B5 Growth and development   109

1 Describe what happens when a shoot is cut off and placed in water. Why is this technique useful?

....

....

.... **[3 marks]**

2 Describe **one** method used to produce a plant clone.

....

....

.... **[3 marks]**

3 Write down **two** functions of auxins in producing plant clones.

....

.... **[2 marks]**

4 Daniel notices that a plant on the kitchen window sill is growing towards the light. Explain how this is of benefit to the plant.

....

....

.... **[3 marks]**

5 Here are some statements about the effects of plant hormones on plant growth. Use the words provided to complete the sentences below.

all directions   **away from**   **one direction**   **photosynthesis**

phototropism   **respiration**   **towards**

When plants are exposed to light from ...., they

grow .... the light. This response of plants to the direction

of light is called .... Responding to light in this way increases

a plant's chances of survival, as light is required for .... **[4 marks]**

6 A scientist investigating the response of plants to light placed one group of plants in the dark, one group exposed to light from one direction, and one group in even illumination. The plants were in an atmosphere of radioactive carbon dioxide, and after five hours, the amount of radioactive auxin in the area below the shoot tip was measured. The scientist's results are shown below.

| | Plants in the dark | Plants in the light | Plants exposed to light from one side | |
			Dark side	Lighted side
Total radioactive auxin in counts per minute	3004	2985	2173	878

Explain fully what these results tell the scientist about the effect of light on auxin in the plants.

....

....

....

.... **[4 marks]**

1 Mitosis is a process that occurs in humans and other organisms.

a Write down a definition of mitosis.

....

.... [3 marks]

b When does mitosis occur in human cells?

....

.... [2 marks]

2 One way of estimating the percentage of time spent during and between mitosis is using the formula below:

$$\text{Percentage of cells in the stage} = \frac{\text{number of cells in the stage}}{\text{total number of cells}} \times 100$$

A student observed 1000 cells taken from different parts of the human gut. Carry out calculations to complete the table below.

Region of human gut	Number of cells in mitosis	Percentage of total time spent in mitosis
Stomach	22	
Small intestine	39	
Large intestine	13	

....

.... [3 marks]

3 The cell cycle is the series of events a) between and leading up to cell division, and b) cell division itself.

The following table gives timings of these two phases.

Cell type	Events leading up to cell division, minutes	Cell division (mitosis)
Meristem of pea plant	1300	150
Chick skin cells	700	25
Rat intestine	2000	30
Developing fruit fly egg	3	7

a What processes are occurring in the events leading up to cell division?

....

.... [3 marks]

b In which of the cells above does mitosis take up the shortest proportion of the cell cycle?

.... [1 mark]

c In which cell type is mitosis longer than the events leading up to it? Suggest why this is.

....

.... [2 marks]

4 Here are some statements about meiosis in humans. Some statements are correct, while some are incorrect.

Put ticks (✓) in the boxes next to the **two** correct statements.

a Meiosis is used to produce eggs and sperm. ☐

b Two daughter cells are produced by meiosis. ☐

c The number of chromosomes in a cell halves during meiosis. ☐

d Meiosis takes place during growth and is used to repair tissues. ☐ [2 marks]

1 Here are some statements about chromosomes, genes and DNA. Use the words provided to complete the sentences.

amino acids   chloroplasts   chromosomes   cytoplasm   DNA   genetic   nucleus

.... are thread-like structures made from a chemical

called ....

This chemical carries the .... code, which codes for the production

of .... in the .... of the cell.   **[5 marks]**

G–E

2 The diagram below is a simplified diagram of a DNA molecule.

D–C

a Label the parts of the molecule shown.   **[3 marks]**

b The DNA molecule is shown flat. Describe how this would appear in three dimensions.

....   **[1 mark]**

3 Here are some statements about protein synthesis. One statement is correct, while the rest are incorrect.

Put a tick (✓) in the box next to the correct statement.

a The genetic code carries the instructions for protein synthesis. ☐

b In the first stages of protein synthesis, RNA is copied in the nucleus. ☐

c Protein synthesis takes place in the nucleus. ☐

d The order of bases in a gene codes for the synthesis of several proteins. ☐   **[1 mark]**

4 The statements below describe how a protein is produced. They are in the wrong order.
Put numbers in the boxes to show the correct sequence. The first one has been done for you.

a In the nucleus, the strands of the DNA molecule separate. ☐ 1

b The mRNA passes from the nucleus to the cytoplasm. ☐

c A molecule of mRNA is synthesised, using the DNA as a template. ☐

d Amino acids are bonded together on a ribosome. ☐

e Amino acids are ferried to the ribosome. ☐   **[3 marks]**

B–A*

B5 Growth and development

1 Here are some statements about the switching on and off of genes in cells.

Some statements are correct, while some are incorrect. Put ticks (✓) in the boxes next to the **two** correct statements.

a A red blood cell makes only the proteins it needs to function. ☐

b In most cells, all the genes are switched on. ☐

c In most cells in the human body, genes are frequently switched on and off. ☐

d In embryonic stem cells, any gene can be switched on. ☐

e Because of their genes, adult stem cells can replace any cell in the body. ☐

[2 marks]

D–C

2 Here are some statements about stem cell research and therapy. Use the words listed below to complete the sentences.

adult   damaged   diseased   embryonic   embryo   limited   many   new

Stem cells have the potential to grow cells to replace .... or

.... cells.

.... stem cells are found at various locations in the body, e.g. the

bone marrow. These cells can be used to produce .... types.   [5 marks]

G–E

3 After a science lesson about stem cells, some friends are discussing the use of stem cells in medicine. Here are some quotes:

Michael: 'Collecting embryonic stem cells from an embryo destroys the embryo in the process. I think using them is therefore unethical.'

Ahmed: 'But most embryonic stem cells come from surplus embryos from IVF treatments. So no harm is being done.'

Beatrice: 'If stem cells can treat conditions that are currently incurable, I think it's OK to produce embryos specifically for therapy.'

Maia: 'I think adult stem cells are just as good, so actually, we don't need to worry about killing any embryos.'

Use these ideas to discuss the use of stem cells in therapy.

....

....

....

....

....

....   [4 marks]

D–C

4 Complete the table below on the use of stem cells from different sources to replace damaged cells of a patient.

Source of stem cells	One advantage	One disadvantage
Embryo		
Adult		
Therapeutic cloning		
Transformed body cells		

[4 marks]

B–A*

Explain what happens when a plant is exposed to light from one side, and why this helps in the plant's survival.

The quality of written communication will be assessed in your answer to this question.

B–A*

[6 marks]

1 Humans have two communication systems.

  a Write down the names of the two communication systems.

.... [2 marks]

  b Write down **two** differences between the systems.

....

....

....

.... [2 marks]

G–E

2 Here are some statements about neurons. Use the words provided to complete the sentences.

central   effectors   eyes   motor   muscles   peripheral

receptors   sensory   stimuli

.... neurons connect ....,

which detect ...., with the ....

nervous system. .... neurons connect the

.... nervous system to ....,

e.g. ...., which produce a response.   **[4 marks]**

D–C

3 Write down the names of **two** hormones. For each hormone, state:

  – where in the body that it is made

  – the effects it has on the body.

....

....

....

....

.... [4 marks]

B–A*

4 The diagram below shows the structure of a neuron.

  a Label the diagram.   [5 marks]

  b On the diagram, draw an arrow showing the direction of the nerve impulse.   [1 mark]

G–E

5 Write down **three** factors that affect the speed of transmission of a nerve impulse. Describe how each factor affects the speed.

....

....

....

.... [3 marks]

B–A*

1 Here are some statements about the way nerves link with other. Some statements are correct, while some are incorrect.

Tick (✓) the boxes next to the **two** correct statements.

G–E

a One nerve can connect physically with many others. ☐

b As a nerve impulse reaches the end of the nerve, a chemical signal is released. ☐

c The junction between one nerve and another is called a synapse. ☐

d Few nerves in the body pass messages to other nerves. ☐ **[2 marks]**

2 The human body is thought to use around 50 different neurotransmitters.

a Give **three** reasons why we need different neurotransmitters.

B–A*

....

....

.... **[3 marks]**

b How are nerves adapted to work with different neurotransmitters?

....

.... **[1 mark]**

3 Here are some statements about nervous co-ordination. Use the words provided to complete the sentences.

axon	brain	ear	effectors	eye
receptors	spinal cord	stimulus	muscle	organ

G–E

The nervous system responds to a change in the environment called a ....

These are detected by special cells called ....

Sometimes these special cells are grouped together or form part of an .... ,

e.g. the .... and the .... **[5 marks]**

4 Here are some statements about nervous co-ordination. Use the words provided to complete the sentences.

central	contraction	enzymes	expansion	heartbeat
hormones	limb movement	peripheral	stimuli	transmitters

D–C

a The part of the nervous system that co-ordinates responses is called

the .... nervous system.

b Glands make and release chemicals such as ....

and ....

c Muscles are used for movement. Their .... helps

the body to move away from or towards .... Muscles are

also used for movement we're not conscious of, e.g. our .... **[3 marks]**

1 If a bright is shone into your eyes, muscles in the iris of your eye contract, reducing the amount of light that enters your eye.

a Complete the flow chart below to show this process.

| Light (stimulus) | → | | → | | → | | → | | → | Muscles in iris (effector) |

[4 marks] G–E

b Explain how this reflex is useful.

....

.... **[1 mark]**

2 You pick up a dinner plate that is hot. The dinner plate is very expensive, and you do not drop it.

Explain how you have prevented yourself from dropping the plate.

....

....

....

.... **[3 marks]**

B–A*

3 A scientist carried out an experiment on the behaviour of woodlice. Twenty woodlice were placed in a choice chamber (see diagram), where four different environmental conditions had been produced.

After one hour, the following results were obtained:

Conditions	Dry		Wet	
	Light	Dark	Light	Dark
Number of woodlice	1	5	3	11

dry light | wet light

dry dark | wet dark

a What percentage of woodlice are found in the light; dark; wet; dry?

....

.... **[4 marks]**

G–E

b What conclusion can be drawn from the experiment?

.... **[2 marks]**

c What type of behaviour are the woodlice showing?

.... **[1 mark]**

d Suggest why this type of behaviour is essential to simple animals.

.... **[2 marks]**

4 The doorbell rings and a person's dog starts to bark loudly. Explain how this is an example of a conditioned reflex.

....

....

.... **[3 marks]**

D–C

5 The hoverfly is a harmless insect that has black and yellow stripes resembling those of a wasp. Explain how a conditioned reflex that develops in predatory birds increases a hoverfly's chances of survival.

....

....

....

.... **[3 marks]**

B–A*

1 The brain co-ordinates the activities of the body.

   a Label the diagram of the brain below.

   b Write down **four** traits, most developed in humans, that the cerebral cortex is most involved with.

   ....

   ....

   ....   **[4 marks]**

2 When investigating how the brain works, explain the advantages of using techniques such as magnetic resonance imaging (MRI) over invasive techniques.

   ....

   ....

   ....

   ....   **[5 marks]**

3 List **four** traits linked with the highly developed structure of our brains that make us human.

   ....

   ....   **[4 marks]**

4 Here are some statements about how we learn things. Use the words provided to complete the sentences.

axons	drugs	gaps	impulses	neuron pathway
neurons	preventing	repeating	links	stimuli

   Transmitting impulses between .... in the brain leads

   to .... forming between the neurons. This is called

   a .... These are strengthened by ....

   the experience, so more and more .... follow the same route.

   Another way of strengthening these is using strong ....   **[3 marks]**

5 Explain why children find it easier to learn new skills than adults.

   ....

   ....   **[1 mark]**

6 Describe and explain what happens if a child is not given the appropriate stimuli early in life.

   ....

   ....

   ....   **[5 marks]**

G–E

D–C

B–A*

G–E

D–C

B–A*

1 Jodie is trying to remember a list of things for her science exam next week.

a Write down a definition of the term **memory**.

.... [2 marks]

b Which type of memory will Ruby need to use?.. [1 mark]

c Write down **two** ways that might help Ruby to remember items in the list.

.... [2 marks]

2 Complete the diagram of the multi-store model of memory opposite.

memory   memory   memory

Forgetting

[5 marks]

3 Some friends are discussing how they are revising for their science exam. Here are some quotes:

Amir: 'When I'm preparing, I condense my science notes into key points.'

Justine: 'When I'm revising a list of points, I use the initial letter of each word, and arrange them into a word or list that I can remember easily. It's called a mnemonic.'

Lucas: 'When I've finished reading through my science notes, I write down as much as possible of what I've read.'

Bethany: 'If I listen to loud rock music while I'm revising, it helps things to sink in.'

a Which of the friends has used a stimulus to help them to memorise their science?

.... [2 marks]

b Which of the friends is using both processes involved in memory? Explain your answer.

....

.... [2 marks]

4 Write down the names of **two** groups of chemicals that interfere with nerve impulses moving between a nerve and another nerve, and a nerve and a muscle.

.... [2 marks]

5 When a transmitter substance called acetylcholine crosses a synapse between a nerve and a muscle, it causes the muscle to contract.

Bungarotoxin, a venom produced by the banded krait snake, blocks acetylcholine receptors. Explain what happens to the muscles of someone bitten by a banded krait.

....

.... [2 marks]

6 One of the effects of the drug MDMA (Ecstasy) is to block the re-uptake of a chemical called serotonin into a neuron at a synapse. Serotonin is a chemical transmitter, which in the brain, is important in regulating mood. Explain the science involved when a nerve impulse is transmitted, and the effect of Ecstasy on this.

....

....

....

....

.... [5 marks]

Use the work of Pavlov to explain how animals can develop and learn a reflex response to a stimulus by introducing a new, unrelated stimulus.

✏ *The quality of written communication will be assessed in your answer to this question.*

B–A*

[6 marks]

1 Mammals are animals that have an internal skeleton.

a What are the **four** main functions of the skeleton?

....

....

....

.... **[4 marks]**

b What has to happen to a muscle so that it can move a bone?

.... **[1 mark]**

2 a Explain what is meant by the term **antagonistic pair**.

....

.... **[2 marks]**

b Why do athletes increase the size of their muscles?

....

.... **[2 marks]**

3 The diagram below shows the muscles and bones in a human arm.

forearm being lowered

biceps

triceps

Explain how this type of joint can enable a person to lift an object.

....

....

....

.... **[4 marks]**

G–E

D–C

B–A*

1 Give **two** examples of a **synovial** joint.

....

.... [2 marks]

2 Match the part of the joint to the correct description.

Synovial membrane	stops end of bone wearing away
Synovial fluid	join ends of bones together
Articular cartilage	reduces friction in the joints
Ligaments	produces synovial fluid

[4 marks]

3 Describe the role and properties of ligaments and tendons in a functioning joint.

....

....

....

.... [4 marks]

4 Suggest **one** advantage that ball and socket joints have over hinge joints.

....

.... [2 marks]

5 What is the advantage of ligaments being elastic?

....

.... [2 marks]

6 Nick is reading a newspaper. An article highlights the risks of different activities giving the chances of being injured in the course of a year. A table from the article appears below.

Activity	Proportion of injuries (per 100 participants per year)
Figure skating	50
Snowboarding	50
Rugby	65
Playing computer games	1
Watching TV	0
Gardening	15

Nick decides to stop his son from playing rugby as he thinks it is safer paying computer games.

Suggest reasons associated with risk as to why Nick may not be making the correct decision.

....

....

....

....

.... [5 marks]

B7 Further biology

1 Triston and Spencer are brothers.
Triston is a computer games programmer. He is 190 cm tall and weighs 115 kg.
Spencer is a boxer. He is 185 cm tall and weighs 100 kg.

a Using the formula for BMI, calculate the BMI of each brother. Show your working.

$$BMI = \frac{body\ mass\ (kg)}{[height\ (m)]^2}$$

....

....

.... **[2 marks]**

b Using the BMI table, state Triston's and Spencer's weight category.

Obese	30+
Overweight	25–29.9
Healthy	18.5–24.9
Underweight	<18.5

....

.... **[1 mark]**

c Using the information given, suggest a reason why Spencer might not be worried about his weight.

....

.... **[2 marks]**

2 Ruth has had an operation on her back. She is overweight. Her physiotherapist has suggested that she undertakes an exercise programme to lose the excess weight.
Suggest why the physiotherapist needs to know her family history and current medication.

....

....

.... **[2 marks]**

3 Why must an exercise programme start slowly and build up in intensity over time?

....

.... **[2 marks]**

4 Graham measures his weight regularly throughout the day. For each weigh-in he always takes three measurements.

a Explain to Graham why measuring his weight so frequently may be a bad idea.

....

.... **[2 marks]**

b Why might the three measurements that Graham takes at each weigh-in be a good idea?

....

.... **[2 marks]**

G–E

D–C

B–A*

1 What is meant by the terms:

  a Heart rate.

.... [1 mark]

  b Blood pressure.

.... [1 mark]

2 What is the word equation for aerobic respiration?

....

.... [2 marks]

3 a Lara is training for the Rio Olympic Games. She is 19. What should her maximum heart rate be? Show your working. [2 marks]

....

....

  b If Lara is to exercise safely, her heart rate needs to be lower than her maximum rate. What is the best percentage range for her heart to be increased to?

....

.... [2 marks]

  c When Lara first starts her training it takes her 8 minutes to recover. Suggest what her recovery time would be if she were much fitter.

.... [1 mark]

G–E

4 Give a definition for each of these skeletal injuries.

  a A sprain.

.... [1 mark]

  b A torn ligament or tendon.

.... [1 mark]

  c A dislocation.

.... [1 mark]

D–C

5 Sprains can be caused by excessive exercise. Explain the correct treatment for a sprain.

....

....

....

....

.... [5 marks]

6 People who are recovering from a skeletal-muscular injury are often referred to a physiotherapist for recovery. Describe what a physiotherapist does.

....

....

....

.... [2 marks]

B–A*

1 Mammals have a **double circulatory system**.

a Which diagram, **A**, **B**, **C** or **D**, shows a double circulatory system?

.... [1 mark]

b Explain the reason for your choice.

....

.... [2 marks]

G–E

2 The heart is a muscular pump. Starting with the vena cava, draw arrows to show the flow of blood through the chambers of the heart for one complete circulation. **[2 marks]**

D–C

3 a Explain why faulty valves in the heart can cause problems with the circulation of blood.

....

....

.... [3 marks]

b Suggest a symptom that someone may present if they had faulty heart valves and explain why you think this would be the case.

....

....

.... [2 marks]

B–A*

1 Which of the following are components of blood? Put a tick (✓) in the boxes next to the correct answers.

hormones	☐	life force	☐
plasma	☐	platelets	☐
red blood cells	☐	white blood cells	☐

[1 mark]

2 The diagram shows a red blood cell.

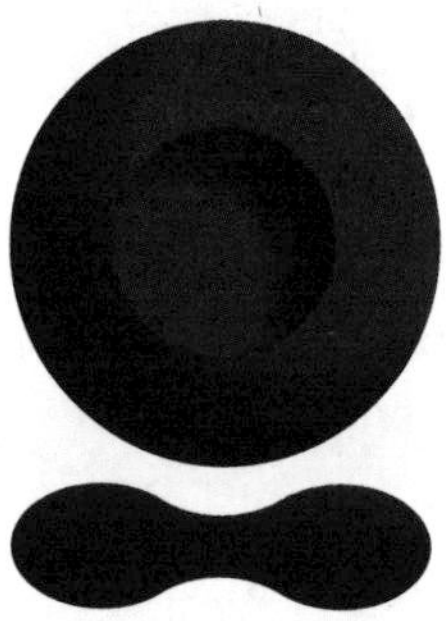

Give a feature, other than its shape, that enables a red blood cell to carry out its role effectively.

.... [1 mark]

3 90% of plasma is made up of water. The remaining 10% is made up of a number of other metabolic compounds.

a Describe the role of salt in the plasma.

.... [1 mark]

b Carbon dioxide and urea are waste products transported in the plasma. Where does each get transported to?

....

.... [2 marks]

c Name **three** other groups of compounds transported in the plasma.

....

.... [3 marks]

4 The diagram below shows a capillary bed.

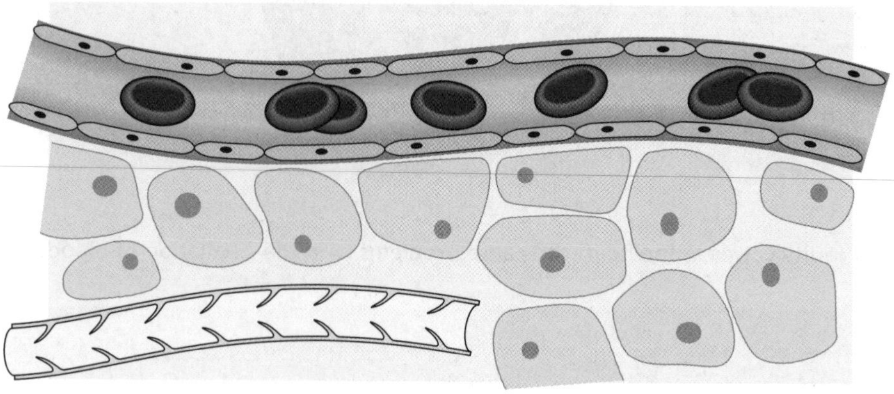

Explain the function of a capillary bed. Make use of specific terminology in your answer.

....

....

....

.... [4 marks]

B7 Further biology

1 Diane is jogging. As she exercises her muscles use up oxygen.

G–E

a What is the name of the chemical in the blood which binds to oxygen?

.... [1 mark]

b The following table shows data of the oxygen and carbon dioxide in Diane's arterial blood and blood returning in veins to the lungs. The labels are missing. Add the labels **artery** and **vein** to the correct parts of the table.

Component		
Oxygen (ml per 100 ml of blood)	5	15
Carbon dioxide (mmHg)	49	35

[1 mark]

2 As well as transporting other chemicals, the blood transports hormones.

a What is the role of **adrenaline** in the body?

.... [1 mark]

D–C

b What is the role of **insulin** in the body?

.... [1 mark]

3 Roz has sickle-cell anaemia.

red blood cell

sickle cell

B–A*

a Explain why the normal red blood cells have their characteristic shape.

....
.... [2 marks]

b Suggest what difference sickle-cell anaemia makes to blood circulation in the body.

....
.... [2 marks]

1 The following stages explain how the body controls its temperature when it gets too warm. They are **not** in the correct order. Put the stages in the correct order by writing the letters in the empty boxes. One has been done for you.

A Receptors in the brain detect the rise in temperature.

B Blood flow to the skin increases.

C Instructions are sent by the brain.

D Body temperature is too high.

E Body temperature returns to the correct level.   [3 marks]

☐ D ☐ ☐ ☐ ☐

2 Sweat contains salts such as potassium and sodium. Why are these needed by the body?

....

....   [2 marks]

3 Ethan is an astronaut carrying out a space walk on the International Space Station. He has to wear a space suit, not only to allow him to breathe but to help control his body temperature.

a Ethan has been working for six hours. He is sweating. Explain why his body produces sweat.

....

....   [2 marks]

b The suit Ethan is wearing collects the sweat and re-processes it into a form that he can drink. Suggest why this is a good idea.

....

....   [2 marks]

4 a Explain what happens during **vasodilation**.

....

....

....

....   [4 marks]

b What would the skin look and feel like in a person who was experiencing vasodilation? Explain your answer.

....

....   [2 marks]

c Explain what happens during **vasoconstriction**.

....

....

....

....   [4 marks]

d What would the skin look and feel like in a person who was experiencing vasoconstriction? Explain your answer.

....

....   [2 marks]

1 The following statements about body temperature are either true or false. Write the letter T in the box for a true statement and an F in a box next to a false statement.

A core body temperature of 37°C is healthy. ☐

A core body temperature of 39°C is not healthy. ☐

A core body temperature of 40°C is not healthy. ☐

A core body temperature of 31°C is healthy. ☐

Core body temperature varies according to the external temperature. ☐

Skin temperature should not drop below 37°C. ☐ **[3 marks]**

G–E

2 When someone is too cold the body responds.

a What are the **two** sources of temperature information from the body used by the brain to decide on a response?

....

.... **[2 marks]**

D–C

b The body can generate heat for a short while if the body is too cold. Give **two** ways that the body achieves this.

....

.... **[2 marks]**

3 George is suffering from hypothermia.

a At what temperature does the core body temperature drop below for it to be classed as being hypothermic?

.... **[1 mark]**

b Explain why it is important not to warm a person who is suffering from hypothermia too quickly.

....

....

....

.... **[4 marks]**

4 a What is the name of the control centre in the brain responsible for monitoring blood temperature?

B–A*

.... **[1 mark]**

b Explain why muscles and sweat glands are antagonistic effectors.

....

....

....

....

.... **[4 marks]**

1 a What is the name of the hormone that is involved in maintaining the correct amount of glucose in the blood?

.... [1 mark]

b One test to see whether a person has diabetes or not is to do a urine test. What is being looked for in the urine?

.... [1 mark]

c Diabetes sufferers can also have problems with blood pressure. This can cause problems elsewhere in the body. Suggest **two** parts of the body that can be affected by high blood pressure.

....

.... [2 marks]

2 Wiremu and Atu both have diabetes. Wiremu has Type 1 diabetes whilst his friend Atu has Type 2 diabetes.

a Explain why Wiremu has to inject insulin.

....

.... [2 marks]

b What is Atu's method for coping with his diabetes? Why is it different to Wiremu's?

....

....

....

.... [2 marks]

c The chart below shows the number of sugary drinks consumed against the relative risk of diabetes.

Describe the correlation between sugar-sweetened drink consumption and risk of Type 2 diabetes.

....

.... [1 mark]

3 Explain in terms of negative feedback how the hormone insulin works.

....

....

....

....

.... [5 marks]

1 Some plastic containers have the following symbol on them:

a What will happen to waste with this symbol?

.... [1 mark]

b What type of system would this product be part of? Circle the correct answer. [1 mark]

closed loop   **open loop**   **waste loop**

c Why is it better to recycle an object? Use the word **resources** in your answer.

....

.... [2 marks]

2 Describe how carbon dioxide is cycled between plants and animals.

....

.... [2 marks]

3 What type of organism uses plant and animal remains as a source of organic matter?

.... [1 mark]

4 The plants and animals whose bodies were used in the formation of crude oil ultimately derived their energy from the Sun.

Explain why the use of crude oil is regarded as being an open loop system and not a closed loop.

....

....

....

....

.... [5 marks]

5 Describe how microorganisms break down waste and explain how this contributes to a closed loop system.

....

....

....

....

.... [5 marks]

1 What are the purposes of a fruit?

....

.... [2 marks]

2 Fruits take a long time to grow.

   a Why do humans and other animals eat fruit?

.... [1 mark]

   b What is the reason why some fruits take a long time to grow?

.... [1 mark]

3 Some fruits, such as the passion fruit, have a high cost. Suggest **two** reasons why a fruit may have a high cost.

....

.... [2 marks]

4 Many organisms produce a large number of reproductive structures.

The cashew nut tree grows in Brazil. It produces thousands of nuts each year.

   a Suggest why this is the case. Explain your answer.

....

....

.... [2 marks]

   b Suggest **two** causes for the nuts failing to grow into an adult tree.

....

.... [2 marks]

5 Broadly speaking a rainforest is an example of a closed loop system. However, there will always be some natural losses.

   a Suggest **two** natural losses in a rainforest.

....

.... [2 marks]

   b Explain how humans are disrupting the cycle.

....

....

....

.... [4 marks]

   c Explain why it is difficult for people living in developed nations to argue that people living in rainforests should not use the forest to improve their lives.

....

....

....

....

.... [4 marks]

G–E

D–C

B–A*

1 Soil is needed for plants to grow. Which of the following are needed in a healthy soil? Tick (✓) the correct boxes.

minerals ☐   air spaces ☐

vitamins ☐   nutrients ☐

water ☐   organic material ☐   **[2 marks]**

2 The growing plants affect the soil in a number of ways. Suggest **two** ways in which plants affect the soil.

....

.... **[2 marks]**

3 The diagram below shows a river running through part of a rainforest.

At points **A**, **B** and **C** samples of water are removed and tested for **turbidity**. Turbidity is a measure of the cloudiness of the water. The cloudier it is, the higher the number.

Sample site	Turbidity (per ml)
A	0.1
B	10.0
C	5.2

a Using the table above, identify the area with the highest turbidity. What is the reason for the increased turbidity?

....

.... **[2 marks]**

b Comparing the site you chose to the one with the lowest turbidity, calculate the percentage increase in turbidity between the two sites.

....

....

.... **[3 marks]**

4 a The United Nations has defined four types of service that an ecosystem provides for the planet. One type of ecological service is provisioning. What are the other **three** types of ecological services?

....

....

.... **[3 marks]**

b Give **two** ways in which humans take advantage of the provisioning service.

....

.... **[2 marks]**

1 Four friends are discussing the environment.

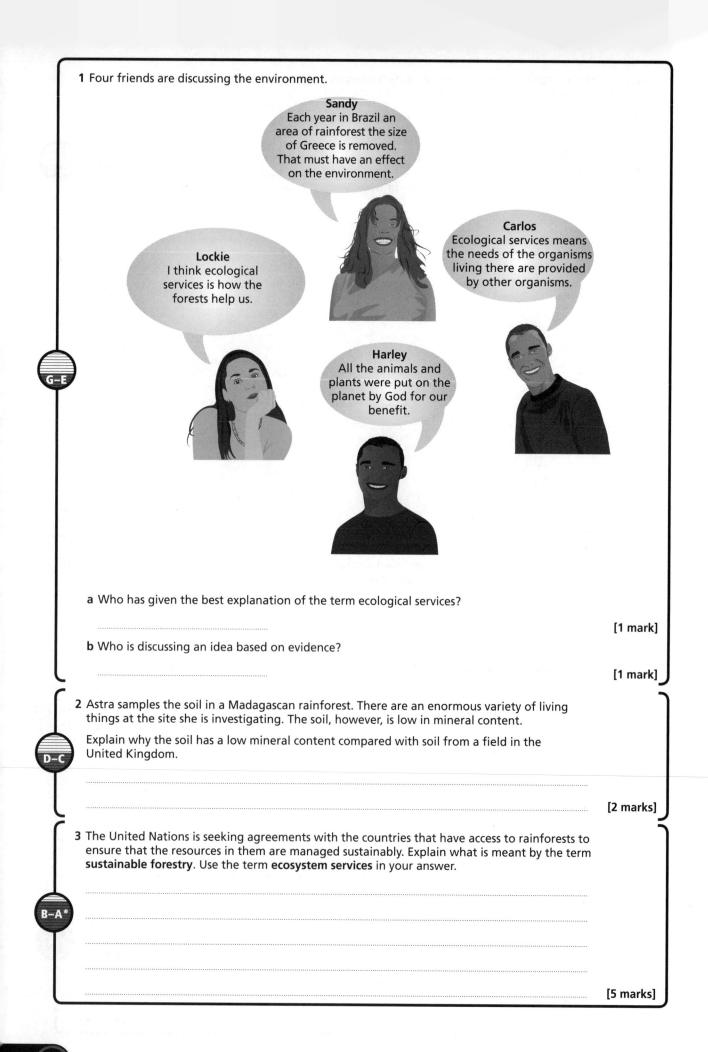

Sandy
Each year in Brazil an area of rainforest the size of Greece is removed. That must have an effect on the environment.

Lockie
I think ecological services is how the forests help us.

Carlos
Ecological services means the needs of the organisms living there are provided by other organisms.

Harley
All the animals and plants were put on the planet by God for our benefit.

a Who has given the best explanation of the term ecological services?

.... [1 mark]

b Who is discussing an idea based on evidence?

.... [1 mark]

2 Astra samples the soil in a Madagascan rainforest. There are an enormous variety of living things at the site she is investigating. The soil, however, is low in mineral content.

Explain why the soil has a low mineral content compared with soil from a field in the United Kingdom.

....

.... [2 marks]

3 The United Nations is seeking agreements with the countries that have access to rainforests to ensure that the resources in them are managed sustainably. Explain what is meant by the term **sustainable forestry**. Use the term **ecosystem services** in your answer.

....

....

....

....

....

.... [5 marks]

1 All waste eventually gets broken down or spread around so much that it is at a non-toxic level in the environment. Many supermarkets in the UK now charge for plastic bags. What is the reason for this? Tick (✓) the best answer.

Supermarkets want to make more money. ☐

The oil to make bags is increasing in cost. ☐

Some bags are collectors items. ☐

Bags take a long time to break down in landfill. ☐ **[1 mark]**

G–E

2 In some parts of the UK, sewage is discharged by pipe into the sea. Which of the following is the best reason for this? Tick (✓) the best answer.

The people in these areas don't know any better. ☐

The population involved is low, so the sea can cope. ☐

It is an expensive way to remove sewage. ☐

There is no other option. ☐ **[1 mark]**

3 a Sewage is high in compounds containing the element nitrogen. What do plants use nitrogen to make?

.... **[1 mark]**

D–C

b If a river becomes depleted in oxygen, which is the only type of respiration possible?

.... **[1 mark]**

4 Read the following article.

> Mercury is a liquid at room temperature. It used to be used in gold mining. The mercury would coat small nuggets of gold, making them easier to see. Some of the gold escaped into the river and was then taken up by small creatures. These were then eaten by larger fish and eventually the fish were caught and eaten by humans. The fish contained very high levels of mercury, causing birth defects and other illnesses in the families eating the affected fish.

a What is the name of the process described in the article by which the concentration of mercury increases throughout the food chain?

.... **[1 mark]**

B–A*

b The diagram shows a food chain and the level of mercury.

water ———▶ zooplankton ———▶ small fish ———▶ large fish ———▶ Human

0.00004 ppm   0.04 ppm   0.4 ppm   40 ppm

Mercury concentration (parts per million)

Calculate the percentage increase in mercury concentration between the zooplankton and large fish. Show your working.

....

....

.... **[2 marks]**

1 In parts of the world large tanks or artificial ponds can be used for fish farming.
The following statements about fish farming using ponds are true or false.
Indicate which is true with a T and which is false with an F. **[2 marks]**

Fish are low in protein, which is why you need a lot. ☐

Fishermen know how many fish are in the pond. ☐

The biomass is the total weight of the pond. ☐

Farmers can use the run off from fields in the pond. ☐

2 In New Zealand in the Marlborough Sounds, a number of salmon farms have been set up.
The salmon are grown in large netted areas of the water.

In 2013 there was a public consultation about whether there should be an increase in the
number of salmon farms.

The Marlborough Sounds is a sheltered waterway that contains a large number of
natural species, such as blue cod, squat lobster, barracuda and gannets.

a Suggest a biological reason that people could have made **against** an increase in the number
of salmon farms.

....

....

.... **[2 marks]**

b Suggest a biological argument **for** a limited increase in salmon farms in the area.

....

.... **[1 mark]**

3 In areas of the world that have fishing quotas there are often size limits on the fish that are
caught. If a fish is smaller than a set size, the fish has to be thrown back.

a Explain why quotas often use a set size of fish as the indicator for keeping or throwing back
a fish.

....

....

.... **[2 marks]**

b What is meant by the term **stock biomass**?

....

.... **[1 mark]**

c Sometimes fish are brought on board which have died of natural, non-fishing-related causes.
Fisherman often view this as a tragic waste as perfectly good fish have to be thrown into the sea.
Suggest why it is illegal to take fish that are smaller than the quota size, even if they were already
dead, even though they were not killed through fishing.

....

....

.... **[1 mark]**

G-E

D-C

B-A*

1 Microorganisms are increasingly used to produce materials for our use. The list below suggests some reasons why microorganisms are used. They are not all correct. Indicate which are true using a T and those that are false using an F.

Microorganisms reproduce very slowly.   ☐

Microorganisms have a simple biochemistry.   ☐

Microorganisms can be genetically engineered.   ☐

There are ethical issues using microorganisms.   ☐

They can only be farmed on a large scale.   ☐   **[3 marks]**

G–E

2 A group of friends is discussing the use of microorganisms in manufacturing.

Bill
All microbes are dangerous, they cause disease. We should leave them well alone.

Ngaire
We can harvest more protein from microorganisms than from animals, such as cows.

Wayne
We already use microorganisms in foods. Stilton is a cheese that is made using *Penicillium* and bacteria.

Diane
Using protein from microorganisms could save the lives of other animals.

D–C

a Who is making an **ethical** statement?

....   **[1 mark]**

b Which **two** people are making an argument based on evidence?

....   **[2 marks]**

3 Fermenters are often used in the manufacture of foods using microorganisms. The microorganisms have to be managed as, if left, they will alter the environment inside the fermenter.

Discuss the typical inputs and outputs from a fermenter containing aerobic bacteria.

....

....

....

....

....   **[5 marks]**

B–A*

B7 Further biology **137**

1 Scientists have been able to take the gene for producing beta carotene out of carrots and put it into rice. Beta carotene also gives the carrot its orange colour. When beta carotene is eaten it is turned into vitamin A by the body. Having rice which can produce beta carotene may save many millions of lives.

a What is the name of the process where a gene is taken from an organism and placed into another, different, organism?

.... **[1 mark]**

b Why is it necessary to introduce the beta carotene gene from the carrot into the rice plant?

....

.... **[1 mark]**

2 Until 1982, insulin could only be produced by extraction from the pancreas of pigs and cows. In 1982 insulin was produced using bacteria expressing the human insulin gene.

a Prior to 1982, an animal rights protestor with Type 1 diabetes could not have refused insulin and survived. Suggest why.

....

.... **[2 marks]**

b Suggest **two** reasons, other than animal rights, why certain groups would have been glad with the insulin produced after 1982.

....

.... **[2 marks]**

3 The following statements are for the sequence involved in using recombinant DNA. They are not all correct. Choose the correct statements and then sort them into the correct order. The first one has been done for you.

A The desired gene is located from a source organism.

B Enzymes cut the gene from the target organism.

C Enzymes cut the gene from the source organism.

D Gene copies are inserted into the vector (a blood cell).

E Gene copies are inserted into the vector (a plasmid).

F The vector is multiplied and inserted into the bacterial cell.

G The vector is multiplied and inserted into the source cell.

H The desired gene is expressed when the cell reaches the end of its cycle.

I The desired gene is expressed when the modified cell develops.

| A | ☐ | ☐ | ☐ | ☐ | **[2 marks]**

1 The process of Fluorescence *In Situ* Hybridisation, or FISH, is used to locate a particular gene in a person's cells.

The process involves four stages.

i) DNA probe preparation.

ii) Fluorescent chemical is attached to the probe.

iii) The DNA probe is mixed with white blood cells and left for a short period.

iv) Cells are washed and then viewed under UV light using a microscope.

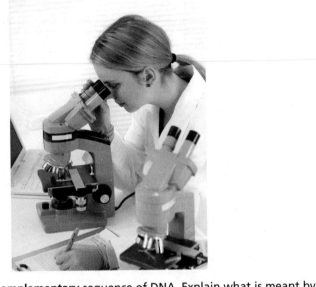

a The DNA probe is made of a **complementary** sequence of DNA. Explain what is meant by the term **complementary**.

....

....

.... **[3 marks]**

b Give the complementary sequence to the following DNA sequence. **[1 mark]**

Target strand: A   G   C   G   T   A

Complementary strand:

c Suggest the reason why the white blood cells, having been left with the DNA probe for a period of time, are then washed before using the UV light and microscope.

....

....

....

.... **[3 marks]**

2 For many conditions there are a variety of genes that contribute to the risk of actually suffering from a genetic disorder.

Explain why it would be ethically wrong for insurance companies to have access to this kind of information.

....

....

....

....

.... **[4 marks]**

1 Nanotechnology is technology that works on a scale of atoms and molecules.

   a What range of length, in nm, does nanotechnology operate over?

   .... [1 mark]

   b DNA molecules are 2 nm wide. A human hair is 80 000 nm wide. Calculate how many widths of DNA would fit into the width of one hair. Show your working.

   ....

   ....

   .... [2 marks]

2 A group of students is discussing nanotechnology.

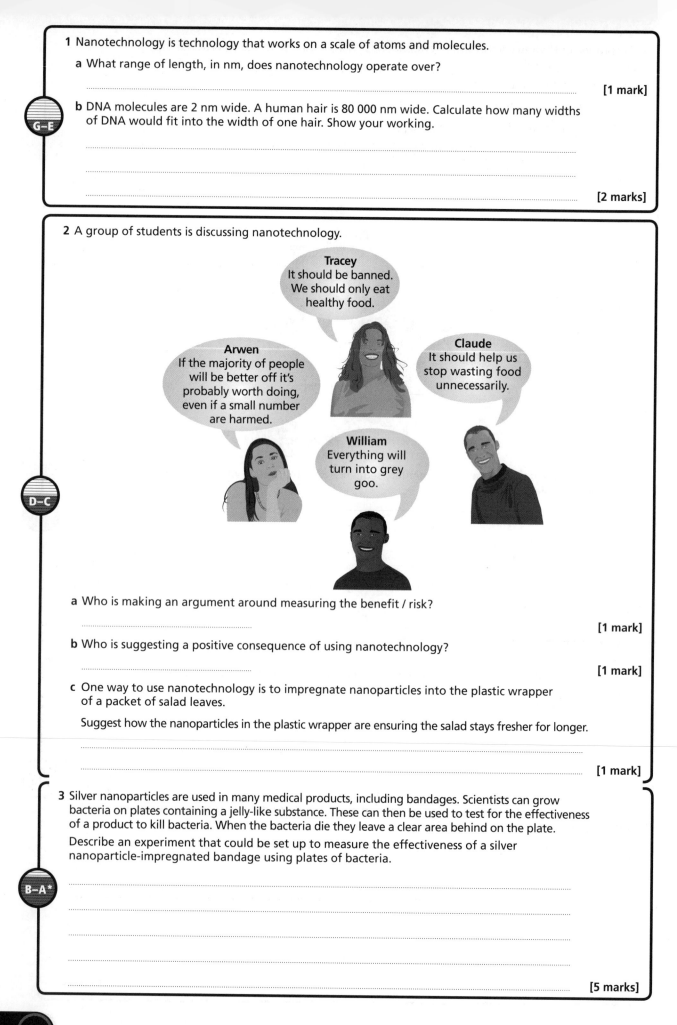

Tracey
It should be banned. We should only eat healthy food.

Arwen
If the majority of people will be better off it's probably worth doing, even if a small number are harmed.

Claude
It should help us stop wasting food unnecessarily.

William
Everything will turn into grey goo.

   a Who is making an argument around measuring the benefit / risk?

   .... [1 mark]

   b Who is suggesting a positive consequence of using nanotechnology?

   .... [1 mark]

   c One way to use nanotechnology is to impregnate nanoparticles into the plastic wrapper of a packet of salad leaves.

   Suggest how the nanoparticles in the plastic wrapper are ensuring the salad stays fresher for longer.

   ....

   .... [1 mark]

3 Silver nanoparticles are used in many medical products, including bandages. Scientists can grow bacteria on plates containing a jelly-like substance. These can then be used to test for the effectiveness of a product to kill bacteria. When the bacteria die they leave a clear area behind on the plate.

Describe an experiment that could be set up to measure the effectiveness of a silver nanoparticle-impregnated bandage using plates of bacteria.

....

....

....

....

.... [5 marks]

Future medicine

1 Gary has been suffering from a heart problem. He has been advised that he needs treatment.

a The following options are discussed. Tick (✓) the option that has the highest risk of tissue rejection.

Replacing the faulty valves with transplanted valves. ☐

Replacing the faulty valves with plastic valves. ☐

Replacing the heart with a transplanted heart. ☐

Replacing the heart with a mechanical pump. ☐ **[1 mark]**

b Suggest **two** reasons why a mechanical heart would save more lives than a transplanted heart.

....

.... **[2 marks]**

2 Stem cells are used to make new specialised cells. They are being investigated to see if they can reverse the paralysis that takes place after the spine is broken.

a Complete the following sentences. Use words from the list below. **[4 marks]**

chemical   differentiated   electrical   energy   signals   undifferentiated

Nerve cells transmit .... through the body. Once a nerve is broken the

.... signals cannot travel past the break. Stem cells are ....

cells. They are being tested in trials to see whether they are effective at reconnecting the nerve cells.

When the stem cells become nerve cells we say they have ....

b IPS cells are created by making adult body cells revert back to being undifferentiated. Explain why they are seen as being an ethical alternative to embryonic stem cells.

....

....

.... **[2 marks]**

3 a Explain how stem cells are used in the treatment of leukaemia, a disease of the blood.

....

....

....

.... **[4 marks]**

b Why is a patient suffering from leukaemia vulnerable to disease, especially immediately after the initial treatment?

....

....

.... **[3 marks]**

G–E

D–C

B–A*

B7 Extended response question

The use of genetic profiling promises to enable treatment to better fit the patient.

In 2013 it was announced that hundreds of genes associated with an increased risk of prostate cancer, breast cancer and bowel cancer had been discovered.

Scientists suggest that everyone should be tested for the presence of the genes that indicate an increased risk of these cancers. Those who carry the high risk genes would be screened earlier and more often for the particular cancer. Those who do not have the genes would not necessarily be screened.

Explain the advantages and disadvantages of using genetic testing before deciding to screen for a cancer.

✏ *The quality of written communication will be assessed in your answer to this question.*

B–A*

[6 marks]

I know that genes carry instructions that control how your body functions.	
I know that our characteristics are controlled by genes and by the environment.	
I understand that genes carry the instructions for us to make proteins.	
I know that we have 23 pairs of chromosomes, and that sex cells have 23 chromosomes (one from each pair).	
I know that the 23rd pair of chromosomes determines our sex (a female is XX; a male XY).	
I understand that a baby has a combination of genes from his or her parents, which leads to variation.	
I am familiar with disorders (Huntingdon's disease; cystic fibrosis) caused by faulty versions of a gene.	
I know that individuals of a clone have identical genes, and clones can occur in nature.	
I know that stem cells are unspecialised cells and are found in the human embryo and adults.	
I can explain that embryonic stem cells can develop into any cell type, but adult stem cells develop into fewer types.	
I am working at grades G/F/E	

I understand that each gene is a section of a molecule called DNA (deoxyribonucleic acid).	
I know that the DNA in our cells is coiled and packed into chromosomes.	
I can explain the difference between structural and functional proteins.	
I know that the different versions of a gene are called alleles.	
I know that for the chromosomes in a pair, the genes for a characteristic are in the same place.	
I understand that across the pair, two alleles can be the same or different.	
I understand how dominant and recessive alleles behave in combination with each other.	
I know how to use a Punnett square and a family tree to show inheritance of characteristics.	
I can describe how genetic testing and screening are used to check people, and embryos, for a disorder.	
I know that stem cells have the potential to treat illnesses.	
I am working at grades D/C	

I know and understand the terms genotype and phenotype, and homozygous and heterozygous.	
I can explain how the sex-determining gene on the Y chromosome triggers the development of testes.	
I know and understand why offspring are both similar and different to their parents.	
I know and understand the implications of testing and selecting embryos before implantation.	
I know that a clone can be produced by transferring an adult body cell to an empty, unfertilised egg.	
I am working at grades B/A/A*	

I understand how microorganisms cause the symptoms of infectious disease.	
I know that in the ideal conditions of the human body, microorganisms can reproduce rapidly.	
I know that white blood cells destroy microorganisms.	
I understand that a vaccine contains a safe form of the microorganism that causes a disease.	
I know that vaccines can never be completely risk-free, and some people show side effects.	
I know that antimicrobials are chemicals used to kill bacteria, fungi and viruses.	
I can explain that an antibiotic is an antimicrobial used to kill bacteria.	
I know that the heart is a double pump.	
I understand about heart rate and blood pressure, and how they can be measured.	
I know that the conditions inside our bodies are kept constant; this is called homeostasis.	
I understand that the kidneys help to balance water in the body by altering the concentration of urine.	
I am working at grades G/F/E	

I understand how to calculate the growth of a population of bacteria.	
I know that microorganisms carry antigens on their surface, and antibodies recognise these.	
I know that bacteria and fungi may become resistant to antimicrobials, and how this can be reduced.	
I understand about the use of placebos when testing new drugs, and the ethical implications of these.	
I know how the structure of arteries, capillaries and veins are related to their functions.	
I understand that high blood pressure increases the risk of heart disease (strokes and heart attacks).	
I know that lifestyle factors (poor diet, stress, cigarettes, drugs) can increase the risk of heart disease.	
I know that body control systems have receptors, processing centres and effectors.	
I am working at grades D/C	

I understand that after an infection, when antibodies have been produced, memory cells stay in the blood.	
I know that to prevent an epidemic, a high percentage of the population must be vaccinated.	
I can explain that antimicrobials are used to inhibit the growth of microorganisms, as well as killing them.	
I know that random mutations lead to microorganisms being less affected by antimicrobials.	
I am familiar with the use of 'open label', 'blind' and 'double blind' testing of new medical treatments.	
I know and understand the importance of long-term trials in investigating the effects of a drug.	
I know that epidemiological and genetics studies help us to understand factors involved in heart disease.	
I know and understand the principle of negative feedback in reversing changes in the body.	
I can explain how water in the body is controlled by ADH, and how drugs (alcohol and Ecstasy) affect this.	
I am working at grades B/A/A*	

I understand that a species is a group of organisms that can breed together and produce fertile offspring.	
I know that the adaptation of a species to its environment means that it can live and reproduce.	
I understand that nearly all species are dependent on energy from the Sun.	
I know that plants absorb only a small percentage of the Sun's energy for photosynthesis.	
I know how environmental change can be monitored using non-living and living indicators.	
I understand that life began on Earth about 3500 million years ago, and the first life was very simple.	
I know and understand the meaning of the terms biodiversity and sustainability.	
I am working at grades G/F/E	

I know that species living in a habitat are dependent on the environment and each other.	
I understand that there may be competition between the plants and animals in a habitat for resources.	
I know about food webs and the impact of removing a species from a food web.	
I know the factors that can lead to the extinction of a species.	
I can explain how energy is transferred between organisms when the organisms or their wastes are fed on.	
I know that energy is lost along a food chain, limiting its length.	
I can calculate the efficiency of energy transfer at different stages of a food chain.	
I know and understand how carbon and nitrogen are recycled in the carbon and nitrogen cycles.	
I know and understand the principles of natural selection (and selective breeding).	
I know that genetic variation (from sexual reproduction and mutation) can be passed on to offspring.	
I understand that evidence for evolution includes the fossil record and DNA analysis.	
I know that maintaining biodiversity is a key to sustainability.	
I am working at grades D/C	

I can explain the interdependence of organisms using food webs.	
I know that energy is transferred when partly decayed organisms are fed on by detritivores.	
I know and understand the processes of nitrogen fixation, protein synthesis and denitrification.	
I can explain how natural selection, with environmental change and isolation, produces new species.	
I understand why Darwin's model of evolution by natural selection is currently the best theory.	
I know that organisms are classified into groups according to similarities and differences.	
I am working at grades B/A/A*	

I know that respiration is the release of energy from food and that it occurs in all living cells.	
I know that respiration provides energy for chemical reactions, growth and movement.	
I understand that photosynthesis uses the Sun's energy; it makes food and energy available to food chains.	
I know the word equations for photosynthesis and respiration.	
I know that enzymes are proteins that speed up chemical reactions.	
I understand that enzymes have an optimum temperature and pH, at which they work best.	
I know how fieldwork techniques are used to investigate the effect of light on plants in the wild.	
I understand that aerobic respiration requires oxygen; anaerobic respiration takes place in the absence of oxygen.	
I know when organisms use aerobic and/or anaerobic respiration.	
I am familiar with the applications of anaerobic respiration.	
I know that the cell membrane regulates what enters and leaves cells.	
I know that oxygen and carbon dioxide move freely in and out of cells by diffusion.	
I am working at grades G/F/E	

I know that enzymes are assembled in the cytoplasm from instructions in genes in the nucleus.	
I know that an enzyme works on the substrate, and one enzyme works with one substrate only.	
I am familiar with the lock and key mechanism of enzyme action.	
I know that temperature increases enzyme activity, up to a certain point.	
I understand that energy for photosynthesis is absorbed by the green chemical chlorophyll in chloroplasts.	
I know that the products of photosynthesis are glucose and oxygen.	
I can explain how the rate of photosynthesis is affected by temperature, carbon dioxide and light intensity.	
I know that diffusion is the movement of molecules from an area of high to low concentration.	
I understand that osmosis is a special form of diffusion involving the movement of water only.	
I know that water is moved into plant roots by osmosis.	
I know that aerobic respiration releases more energy than anaerobic respiration.	
I can explain differences in cell structure between animals, plants, yeast and bacteria.	
I am working at grades D/C	

I understand how high temperatures and extremes of pH can denature enzymes.	
I know that active transport uses energy and is needed by plant roots to absorb nitrates.	
I know the symbol equations for photosynthesis and respiration.	
I am working at grades B/A/A*	

I know that in multicellular organisms, cells become specialised to do particular jobs.	
I know that in animals and plants, cells are grouped into tissues, and tissues into organs.	
I understand that sections from plant stems called cuttings will develop into new plants.	
I know that a type of cell division called mitosis is responsible for growth and repair.	
I understand that when a cell divides by mitosis, two identical daughter cells are produced.	
I know that a type of cell division called meiosis is used to produce gametes (sex cells).	
I know that chromosomes are found in the nucleus, and each is made from a DNA molecule.	
I understand that proteins are assembled in the cytoplasm, instructed by the genetic code in the nucleus.	
I know that only the genes required for a cell to carry out its function are switched on.	
I am working at grades G/F/E	

I know that a zygote (fertilised egg) divides by mitosis to form an embryo.	
I know that the cells in an embryo at the eight-cell stage are embryonic stem cells.	
I know that embryonic stem cells are unspecialised, and can become any type of specialised cell.	
I understand that adult stem cells can develop into certain cell types.	
I know that in plants, the only cell division by mitosis occurs in regions called meristems.	
I know that plants from cuttings are clones of the parent and will have identical features.	
I understand that a plant grows towards the light, and this response is called phototropism.	
I know how growing towards the light will increase a plant's chances of survival.	
I know and understand why, in gamete production, the chromosome number of the cell is halved.	
I understand the structure of DNA (a double helix) and how the four bases pair up.	
I understand that in embryonic stem cells, any gene can be switched on to produce any specialised cell.	
I know that embryonic and adult stem cells have the potential to repair damaged tissues.	
I know that because of ethical issues, the use of embryonic stem cells is government-regulated.	
I am working at grades D/C	

I understand that the growth of roots on plant cuttings can be promoted by plant hormones called auxins.	
I know that auxin is redistributed, and the effects of this, when a plant is lit from one side.	
I understand the events during the growth phase and mitosis phase of the cell cycle.	
I know that the order of bases in a gene defines the order of amino acids assembled into a protein.	
I know that a copy of the gene is carried into the cytoplasm as messenger RNA.	
I know that it is now possible to switch on genes in some body cells to produce required cell types.	
I am working at grades B/A/A*	

I understand the definition of a stimulus, and that stimuli are detected by receptors.	
I know that the nervous and hormonal systems coordinate our responses to stimuli.	
I know that the cerebral cortex is connected with traits that make us human.	
I know that nerve cells, or neurons, transmit electrical impulses when stimulated.	
I am familiar with the structure of a neuron (cell membrane; cytoplasm; nucleus; an extension called an axon).	
I know the path followed by a nerve impulse in a reflex arc.	
I know that reflex actions enable 'automatic' responses to aid survival, and can give examples.	
I understand that nerve impulses are transmitted across gaps between nerves called synapses.	
I know that chemical transmitter substances transmit an impulse across a synapse.	
I know that behaviours in simple animals are instinctive and depend on reflexes.	
I know that as humans interact with their environment, new neuron pathways are formed.	
I understand that memory is the storage and retrieval of information, and that there are two forms of memory.	
I am working at grades G/F/E	

I know that the central nervous system (CNS) is made up of the brain and spinal cord.	
I know that the peripheral nervous system (PNS) is made up of the nerves.	
I understand that the axon of a nerve is covered with a fatty (myelin) sheath, which has gaps.	
I am familiar with how conditioning works and can give two examples of conditioned reflexes.	
I am familiar with the techniques used to map the brain.	
I know that neuron pathways can be strengthened by repetition.	
I understand how models can be used to describe memory, including the multi-store model.	
I am working at grades D/C	

I know about and can compare the responses of the nervous and hormonal systems.	
I know that the myelin sheath insulates the nerve and speeds up the transmission of nerve impulses.	
I know that a transmitter binds to a receptor and initiates the nerve impulse in a second nerve.	
I understand that some drugs affect the transmission of nerve impulses across a synapse.	
I know how Ecstasy affects the concentration of a transmitter called serotonin.	
I know how, in certain circumstances, a reflex action can be overridden.	
I know that conditioning can develop in response to a new stimulus, introduced with the primary stimulus.	
I understand that because of the huge number of potential neuron pathways, humans are able to adapt.	
I know about evidence to suggest that children may only acquire skills at a certain age.	
I can explain how models can be used to describe memory, and their limitations.	
I am working at grades B/A/A*	

I know that for a bone to be lifted muscles must contract.	
I know the names for the different parts of the heart and the components of blood.	
I know that the core body temperature is close to 37°C.	
I know that insulin controls blood sugar levels.	
I can recognise an open and closed loop system.	
I know why there is a lot of energy stored in fruits.	
I know the names of the four ecosystem services.	
I know that waste is sometimes put into the sea and that, as long as it is not overloading the system, this can be handled by the system.	
I know that fish are a source of protein and that we can farm them.	
I know the reasons why bacteria are used to produce materials.	
I know that genes can be removed from one organism and be transferred to another.	
I am working at grades G/F/E	

I know that for a bone to be moved muscles contract and relax in pairs.	
I can describe the route blood flows through the heart.	
I know how the body responds to cold and excess heat.	
I know that there are two types of diabetes and the treatments for each.	
I can describe simple cycles, such as for carbon dioxide, between plants and animals.	
I know why plants and animals produce large numbers of reproductive structures, such as fruit.	
I can explain why the soil in the rainforest has less nutrients than in a temperate location.	
I know that when too much waste enters a habitat, such as a pond, the nitrogen in the waste can cause the plants and bacteria to grow too much.	
I can suggest reasons why fishermen have quotas on the fish they catch.	
I know why bacteria can be made to produce insulin.	
I am working at grades D/C	

I can explain how muscles operate antagonistically to one another.	
I can explain why mammals have a double circulatory system.	
I can explain how the body responds to hypothermia and heat exhaustion.	
I can explain how the body manages blood sugar levels in terms of negative feedback.	
I can explain how the formation of crude oil is not a closed loop system.	
I can explain how human activity can change the ecosystem services.	
I can explain the terms eutrophication and bioaccumulation.	
I know the sequence of events involved in genetic modification.	
I can explain how a gene can be located and identified in cells.	
I can explain the issues behind the use of nanotechnology.	
I am working at grades B/A/A*	

Notes

B1 You and your genes

Page 84 What genes do

1 a Nucleus

  b DNA (deoxyribonucleic acid)

2 a Collagen or keratin

  b Enzyme, e.g. amylase or antibody or hormone

3 a The project identified the location of genes on the different chromosomes, enabling scientists to improve their understanding of the way in which genes control our development and function; OR providing scientists with an opportunity of understanding diseases that have a genetic cause OR diseases where genes provide protection against the disease, and an opportunity to control or screen for the disease

  b In identifying genes linked with disease: some drugs' companies have tried to patent/claim ownership of genes, limiting research; OR some insurance companies might withhold insurance from someone likely to get a genetic disease

4 a Blue eyes; dimples; straight hair (Any 2)

  b Pink hair; decayed tooth; pierced ear; scar (Any 2)

  c Skin colour. Skin colour is controlled by genes, but is affected by exposure to the Sun (an environmental factor)

5 Eye colour or height; skin colour/height will show a continuous range across the population

6 Genotype refers to the genetic description of the person for the characteristic, written as letters, e.g. for a person with dimples, DD. Phenotype is a description of the features of the organism, e.g. for dimples, has dimples or does not have dimples. The phenotype will depend on the person's genes, but may also be affected by how these interact with the environment

7 c

Page 85 Genes working together and variation

1 a; c

2 a A mistake occurs when producing sex cells; so that the individual has a different number of chromosomes to the normal 46

  b Down's syndrome (or another suitable example); the person has an extra chromosome 21/ three chromosome 21s instead of two

3 You are similar because: you inherit genes from your parents (half from your mother and half from your father); these genes control your characteristics; some behavioural characteristics will be learned from parents

  You are different because: sex cells contain half the genetic material from each parent; combinations of genes/ chromosomes from each parent will make you different; you may inherit genes not seen (expressed) in either parent; the environment will also produce differences, e.g. diet and exercise

4 Different forms of a gene

5 The term homozygous means that both alleles for a characteristic are the same (on each chromosome of the pair). So the combinations DD and dd are homozygous. The term heterozygous means that the alleles for a characteristic are different (on each chromosome of the pair). So the combination Dd is heterozygous

Page 86 Genetics crosses and sex determination

1 Identical positions; each chromosome; alleles

2 a Dominant; the genotype is written in upper case letters

  b Tt; tt

  c The phenotype for Tt is tongue roller. The phenotype for tt is non tongue roller (cannot roll tongue)

3 a The couple could be TT; or Tt

  b i The genotype of both the man and women must be Tt; without the presence of a t allele in both parents, all the children would be tongue rollers

   ii

		Mother possible alleles in eggs	
		T	t
Father possible alleles in sperm	T	TT Tongue roller	Tt Tongue roller
	t	Tt Tongue roller	tt Non tongue roller

In the Punnett square:
- 1 mark for the different alleles that could be passed on to the offspring from the mother and father (the alleles in the egg cells and sperm cells)
- 1 mark for the possible combinations (genotypes)
- 1 mark for the phenotypes produced

4 The presence of the sex determining gene on the Y chromosome in the cells of the embryo means that it will be male. It is thought that the sex-determining gene triggers the development of testes in the embryo

5 The genes for some traits such as haemophilia and colour blindness are found on part of the X-chromosome not present in the Y-chromosome. If a child inherits a defective allele for blood clotting or normal colour vision, a male baby will have the condition because they have just a single X-chromosome. But for a female baby inheriting a defective allele, it's likely that a normal allele is present on their other X-chromosome. Although there is a small chance that a female baby will inherit two recessive genes from their parents

Page 87 Gene disorders, carriers and genetic testing

1 a Dominant; Huntingdon's disease; concentrating

  b Recessive; cystic fibrosis; digesting food

2 Write out your working stage by stage.

  There are 700 000 babies born.

  So, if there are 1 in every 10 000 babies born with the allele, there are

  $\dfrac{700\,000}{10\,000}$ in every 700 000 babies born

  = 70 born in the UK every year

  Or: the proportion of babies born = $\dfrac{1}{10\,000}$ = 0.00001

  So, if there are 700 000 babies born each year, the number born

  = 0.00001 × 700 000 = 70

3 a So that people can get treatment for the disorder (if it is treatable); allow people to plan for the future (if treatment of the disorder is less successful, or it is untreatable)

  b Amniocentesis; Chorionic villus sampling

4 a i Used to detect five different conditions, so identification; may make treatment possible

    ii Has ethical implications, because identification of a disorder may cause employer to change work role; an insurance company to withhold insurance

  b Embryo testing detects genetic disorders; only embryos free of genetic disorders are then implanted

Page 88 Cloning and stem cells

1 a Successful characteristics can be passed to offspring; advantage when plants or animals live in isolation (Any 1)

  b As they are genetically identical; changes in conditions could result in the population being wiped out

2 The nucleus from a body cell is extracted; and inserted into an egg cell; that has had its nucleus removed (giving the egg cell a full set of genes without having been fertilised); the embryo is implanted into a suitable surrogate mother

3 Embryonic; embryo; any; unspecialised; diseases; certain

4 a Embryonic stem cells are found in the five-day-old embryo. They have the potential to develop into any cell in the human body. Adult stem cells are found in adults in a few places in their bodies. They are able to develop into fewer cell types than adult stem cells

  b Embryonic stem cells are extracted from embryos. In the process, the embryo is destroyed. There is no destruction of an embryo / potential human life / ethical implications in the use of cells from a person's skin

5 Stem cells are unspecialised so can be used to produce different types of body cells. They could be used in the testing of new drugs. Understanding how cells become specialised in the early stages of the person's development by the switching on and off of particular genes. Renewing damaged or destroyed cells in spinal injuries, heart disease, Alzheimer's disease and Parkinson's disease

Page 89 B1 Extended response question

5–6 marks

Explains why the genotypes of the parents must be Pp, the cross involved and produces a suitably detailed and accurate Punnett square. Explains that there will always be a 1 in 4 (25%) chance of producing a baby with PKU because which 'type' of sperm produces which 'type' of egg will always be random. All information in answer is relevant, clear, organised and presented in a structured and coherent format. Specialist terms are used appropriately. There are few, if any, errors in grammar, punctuation and spelling

3–4 marks

States that the parents are carriers, and draws an incomplete or not fully-detailed Punnett square for the cross. States that fertilisation is a random event but does not fully explain how this relates to probability of the offspring having PKU. For the most part the information is relevant and presented in a structured and coherent format. Specialist terms are used for the most part appropriately. There are occasional errors in grammar, punctuation and spelling

1–2 marks

Limited description of the cross. Little or no explanation of why the chance of producing a child with PKU will always be the same for the couple. Answer may be simplistic. There may be limited use of specialist terms. Errors of grammar, punctuation and spelling prevent communication of the science

0 marks

Insufficient or irrelevant science. Answer not worthy of credit

B2 Keeping healthy

Page 90 Microbes and disease

1 Bacteria make us feel ill by releasing toxins. Microorganisms that cause disease are called pathogens

2 The number of bacteria after 3 hours would be 2 621 440. So the area of flesh eaten would be 2.62 cm^2

3 a Lag phase; exponential (logarithmic) phase; stationary phase; death phase

   b Phase 1: Lag phase – the bacteria are not reproducing. They are copying DNA and producing proteins

   Phase 2: Exponential (logarithmic) phase – the bacteria are rapidly growing and dividing

   Phase 3: Stationary phase – food is running out. The bacteria are dying at the same rate as they are reproducing

   Phase 4: Death phase – the bacteria are dying as toxins build up

4 Skin; oil; saliva; tears; stomach; acid

Page 91 Vaccination

1 Memory cells are left in the blood after vaccination. An infecting microorganism is destroyed very quickly by antigens in the blood

2 a Builds up immunity

   b To protect all children against diseases that are preventable

   c Antigens on the surface of the flu virus change over time/do not stay the same, giving new strains. Antibodies produced after exposure to one vaccine would not work against new antigens. New vaccines are produced regularly from different strains of the virus

3 William is correct in that the widespread use of vaccines has already eradicated one disease from the world – smallpox

   In response to Xavier's point, although it may be impossible to vaccinate the whole of the world's population, a high percentage would do

   Yvonne is incorrect. Vaccines are very safe. But some people would be reluctant to have their children vaccinated, as there is a very small chance of them causing harm

   Zak is correct. We do not yet have vaccines for all infectious diseases. Also, some microorganisms, such as the flu virus, change continuously, so flu would be difficult to eradicate completely using a vaccination programme

4 a Side effect

   b Because of genetic variation

5 Effective antimicrobials either kill bacteria; or inhibit their growth

Page 92 Safe protection from disease

1 a The normal population of bacteria contains some bacteria that are resistant to the antibiotic. These survive the treatment with the antibiotic. They reproduce (by splitting into two/binary fission). The genes for antibiotic resistance are passed on. The genes for antibiotic resistance spread through the population

   b Reduce their use/only prescribe when absolutely necessary. Complete the course. Never pass on antibiotics prescribed for you to anyone else

2 a The resistance of the bacterium to three antibiotics increases; from 1985/6 to 2000 (for penicillin) and 2001 for erythromycin and tetracycline; then shows a fall. Almost no resistance was found in the bacterium to ofloxacin until 1995. Resistance since has been more or less stable, at around 1%

   b There is a decrease in total antibiotic use between 1997 and 2007

   c There is a correlation between the decrease in antibiotic resistance from 2001 and the decrease in antibiotic use. But this does not prove the reduction of antibiotic use was the cause

3 One type is carried out on healthy volunteers, to check for safety. A second type is carried out on people with the illness, to test for safety and effectiveness

4 In the double-blind test, neither patient nor researcher knows who is receiving the new drug, existing drug or placebo. That way, researchers can have more confidence that any difference between the new treatment/existing treatment/placebo is real. Also, it means that the research is not affected by perceptions of doctors, patients or data analysts

Page 93 The heart

1 a Oxygen; nutrients; wastes

   b Double pump; half; lungs

   c Heart; blood vessels; blood

2 Arteries transport blood away from the heart under high pressure; walls are very thick, elastic and muscular to withstand the pressure

   Capillaries link arteries and veins; walls are one-cell thick to allow the transfer of substances to and from cells

   Veins collect blood and return it to the heart; walls contain elastic, muscular tissue, but are thinner than those of arteries; the blood is under low pressure and veins have valves to prevent the backflow of blood

3 Build-up of fatty material in the coronary arteries, reducing or preventing blood flow. These are arteries that supply the heart muscle with oxygen

4 a There is a fall in deaths from CHD in all age groups; in the early and late 1970s, the death rate rose in the 45–54 and 35–44 age groups

   b The 45–54 age group

   c Smoking decreased; more exercise; improvement in diet / less obesity / less saturated fat in diet (Change in lifestyle = 1 mark only)

   d Review statistics / carry out research on smoking, exercise and dietary habits

Page 94 Cardiovascular fitness

1 Stroke/ heart attack; damage to the organs, e.g. the kidneys

2 Stress; poor diet; smoking; misuse of drugs (Any 3)

3 Pulse rate is measured in beats per minute. Blood pressure is measured using a sphygmomanometer

4 a i In nurses not exposed to cigarette smoke = 17

     ii In nurses exposed to any amount of cigarette smoke = 135

   b How many times more likely to get CHD were the nurses who were exposed to smoke, compared with those exposed to no smoke = 7.9 / 8 times

   c The data suggest that occasional exposure to cigarette smoke is more likely to cause CHD than regular exposure. With regular exposure to cigarette smoke, there is a greater chance of the CHD being fatal: a 1 in 4.5 chance (or 22%) of the CHD being fatal, compared with 1 in 5.7 chance (or 18%) for occasional exposure

   d Studies are difficult to carry out on large samples/numbers of volunteers. Scientists can't expose large numbers of volunteers to cigarette smoke. If the study is carried out on a normal sample, people in the sample do other things in their lifestyles that might affect the result/it is difficult to control other factors

Page 95 Keeping things constant

1 The process by which internal factors in the body are kept constant

2 Temperature

3 Flow is from: receptor; to processing centre; to effector

4 a Hypothalamus; effectors cause vasodilation; heat lost from skin surface; body temperature falls; receptors detect body temperature

   b The receptors detect this and pass the message to the processing centre. The effector (the blood vessels) return to their normal size

   c Negative feedback

5 a Blood plasma is diluted; the kidneys produce a large volume of urine

   b Blood plasma becomes concentrated; the kidneys produce a small volume of urine

6 a ADH acts on the kidneys and causes them to reduce the amount of water lost in urine

   b Alcohol suppresses the release of ADH. ADH increases the amount of urine produced. Alcohol therefore causes us to produce a greater volume of more dilute urine

Page 96 B2 Extended response question

5–6 marks

Provides a detailed discussion responding to the points in the article, referring to the data in the Vaccination Green Book and explaining how this refutes the article's claims:

- The number of deaths from the different diseases was not negligible in the year before the vaccine was introduced. In 1938, over 2000 people in the UK died from diphtheria.
- Introduction of the vaccine caused a marked decrease in numbers of deaths / reduced deaths from diphtheria, measles and polio to zero in 2003. However, in Hib, the number of deaths was still 58% of what they were before it was introduced in 1991. But we do not know the percentage of children vaccinated for Hib.

Answers

- The article refers to 'waves of epidemics' and 'many deaths'. We cannot comment on this, because the table does not provide information on the years immediately after the vaccine – just 2003 data. And we do not have data on other diseases; just six (although these are some of the most serious).

Overall, the data suggest that vaccinations are beneficial, and not harmful, although we do know that a very small minority of people are affected severely by vaccines. All information in answer is relevant, clear, organised and presented in a structured and coherent format. Specialist terms are used appropriately. There are few, if any, errors in grammar, punctuation and spelling

3–4 marks
There is some discussion of the number of deaths from the different diseases prior to their introduction, and the recognition that these were not negligible in the year before the vaccine was introduced. There is comment on the reduction of deaths after vaccination, and the lesser lack of effectiveness against Hib. Limited discussion, or just one point, regarding the limitations of the data, e.g. data only covers six diseases. For the most part the information is relevant and presented in a structured and coherent format. Specialist terms are used for the most part appropriately. There are occasional errors in grammar, punctuation and spelling

1–2 marks
Limited discussion of patterns in data, but concludes that vaccinations have reduced deaths. Answer may be simplistic. There may be limited use of specialist terms. Errors of grammar, punctuation and spelling prevent communication of the science

0 marks
Insufficient or irrelevant science. Answer not worthy of credit

B3 Life on Earth

Page 97 Species adaptation, changes, chains of life

1 a A species is a group of organisms that are able to breed together and produce fertile offspring

  b Ability to accumulate and store water in the stem; thick, waxy cuticle to reduce water loss over the surface; spines instead of leaves to reduce water loss; a stumpy shape that gives a small surface area in relation to its volume; a shallow, spreading root system that stores water when it rains (Any 3)

2 i The foxes feed on shrews, whose population will decrease because of the lack of insects and spiders. The foxes also eat rabbits, however, so the population of foxes is unlikely to be affected, unless the population of rabbits is small

  ii The insects and spiders form the main food for the birds, so food will be in very short supply. As the birds can fly, it is likely that they will move to another area where the food is more plentiful

  iii The insects and spiders form the main food for the lizards, so food will be in very short supply. The lizards are likely to starve and die. They are less likely than the birds to be able to move to another area where the food is more plentiful

  iv The insect-eating birds will either move to another area or die. The owls will have to hunt for food in another area to survive

3 A change in the environment, or an example of a change in the environment, e.g. loss of habitat, climate change, predated on by newly-introduced species, or disease

4 a 15% (111 − 94/111 × 100 = 17/111 × 100 = 15%)

  b The energy is transferred to decomposers as they feed. Decomposers and detritivores use the grasshoppers' bodies to obtain their energy by respiration; some of this energy is lost as heat

Page 98 Nutrient cycles, environmental indicators

1 a Carbon dioxide; air; photosynthesis; fixed

  b Respiration; decomposition; bacteria; fungi; soil

2 Nitrogen-fixing involves bacteria (in the soil or in root nodules) converting atmospheric nitrogen into nitrates; denitrification involves converting nitrates and other nitrogen-containing compounds into atmospheric nitrogen

3 Mayfly larvae are good biological indicators because they can only live in clean water. Monitoring carbon dioxide gives information on climate change

4 a Carbon dioxide levels in the air show a steady increase from around 315 ppm in 1958 to 370 ppm in 2010. The increases are the result of pollution from the burning of fossil fuels. But this does not give a full explanation of the data. The graph also shows yearly peaks and troughs. These seasonal changes are the result of plant activities, and the removal of carbon dioxide from the air for photosynthesis

  b Carbon dioxide in water (the oceans and freshwater); carbon dioxide in air trapped in polar region ice (ice core data)

Page 99 Variation and selection

1 Million; simple; complex; evolution; variation

2 Fossils of eggs that have been found are where the eggs have been turned into rock

3 A mutation is the change in the genetic information in a cell that can occur at random; a mutation will result in a change in the characteristics of an organism; if a mutation occurs in a sex cell, the characteristic will be passed on to the next generation; most mutations are harmful, but those that give the individual an advantage will be passed on, and lead to changes in the population as the gene spreads through it

4 Sometimes genes change because of mutations. Most mutations are harmful, but some will be beneficial. The beneficial changes will spread throughout the population. So the gene pool – the alleles of genes that occur in a population of organisms – will change; some alleles will become more common, others will disappear. Over millions of years, with the change in frequencies of alleles, a new species will arise

5 The cattle breeder selects a female animal/cow that produces milk with a low fat content (but also large volumes of milk) and breeds this with a male. (Be careful! Usually in selective breeding, the scientist would select a male and female with the best qualities, but only females produce milk!) The cattle breeder breeds these from the offspring, then selects the female that produces the milk with the lowest fat and largest volume. The selective breeding process is continued over several generations

6 a In a population, some rats will be resistant to the rat poisons and survive their use. These rats will pass on their resistant genes. These will spread through the population as the rats breed, until most of the rats become resistant

  b Stadtlohn. This town has the highest percentage of poison-resistant rats, because the rat poison has been overused

Page 100 Evolution, fossils and DNA

1 b, a, d, c, e

2 The Galapagos Islands are volcanic, so Darwin realised that ancestors of the organisms on the islands must have arrived, at some point, from the mainland. He observed that the organisms there, such as mockingbirds, were similar but had slight differences to those on the mainland. The mockingbirds were also different from one island to the next. This led to Darwin's idea that species were not fixed, but could change over time

3 Classified; groups; appearance; DNA; evolved; ancestor

4 Fossil evidence: shows that the simplest organisms are found in the earliest rocks, with more complex ones appearing in younger rocks; more recent fossils have features that look like adaptations or developments of those of older organisms. DNA analysis: of today's organisms has confirmed predictions made from the fossil record, including when branches in the tree of life occurred, and how long ago common ancestors were shared and these branches occurred

5 Characteristics that change during the lifetime of an organism are not usually passed on to the offspring unless they occur by mutation in the sex cells; so changes to a giraffe's neck during its lifetime would not be passed on to the next generation; Darwin suggested that new species of organisms evolved by natural selection; differences in the inherited features of organisms that were advantageous would be passed on; over time, a new species would evolve

Page 101 Biodiversity and sustainability

1 A plantation of tropical oil palm trees has very high biodiversity

2 Meeting our current needs without depriving future generations. Recycling and re-using items instead of throwing them away

3 Each organism in an ecosystem is interdependent/dependent on each other. One organism feeds on another, and will itself be food for another. The removal of one organism will mean that others starve, unless they have another source of food, and may mean that the numbers of others, that are preyed upon, increase

4 a Biodiversity is reduced by: growing only one crop in the field; removing hedgerows (along with the organisms that live in the hedgerows); spraying herbicides and pesticides that kill plants and animals

  b Replanting hedgerows; creating beetle banks

5 Use of less, and sustainable packaging materials; considering the materials used; considering the energy used; considering the pollution created (Any 3)

6 They still use energy in their transport; they will decompose slowly in oxygen-deficient landfill sites

Page 102 B3 Extended response question

5–6 marks

Carries out thorough analysis of the data. Appreciates that yields are in tonnes per hectare, and number of plants per hectare varies, so calculates average yields of bananas per plant (Plot 1 – 13.8 kg per plant; Plot 2 – 12.0 kg per plant; Plot 3 – 10.4 kg per plant). Describes trends across plots 1–3: average height of plants increases from plot 1 to 2 (by one-fifth); there is no further increase from plots 2 to 3; time to reach harvest increases; average mass of a bunch of bananas decreases; total yield of bananas per plot increases; total yield of bananas per plant decreases. Gives detailed suggestions for differences: the plants are competing for resources, including light, water and minerals; as the density of plants in the field increases, the availability of these resources to each plant decreases; so the growth/ yields of each plant is reduced. Makes a suggestion for the plants' increased height, e.g. the plants are growing taller to try to 'reach'/'capture' light. All information in answer is relevant, clear, organised and presented in a structured and coherent format. Specialist terms are used appropriately. There are few, if any, errors in grammar, punctuation and spelling

3–4 marks

Reasonable analysis of the data, but no or little in the way of mathematical comparison. States that there is a reduction in growth of the banana plants, and suggests that competition is a factor in this reduction. For the most part the information is relevant and presented in a structured and coherent format. Specialist terms are used for the most part appropriately. There are occasional errors in grammar, punctuation and spelling

1–2 marks

Makes simple comments and limited or no suggestions for the trends in data. Answer may be simplistic. There may be limited use of specialist terms. Errors of grammar, punctuation and spelling prevent communication of the science

0 marks

Insufficient or irrelevant science. Answer not worthy of credit

B4 The processes of life

Page 103 The chemical reactions of living things

1 c, d

2 A chemical that speeds up the rate of a chemical reaction (but is, itself, unchanged)

3 Proteins; amino acids; genes; substrate; active site; lock and key

4 a Stomach: pH 1.5; small intestine pH 8.0

  b Enzyme activity depends on the substrate (protein) being able to fit into the active site of the enzyme. Each enzyme has an optimum pH. At other pHs, the structure of the active site changes, and the substrate is unable to fit as well. At a certain point, the structure of the enzyme is changed permanently (it is denatured) and is no longer able to work

5 a The graph only shows that papain is unaffected by pH between 4.0 and 8.5. We do not know how its activity is affected at lower or higher pHs

  b The shape of the active site of the enzyme is unaffected by pHs from 4.0 to 8.5

Page 104 How do plants make food?

1 Reactants are carbon dioxide and water; products are oxygen and glucose. The reaction is driven by light energy – in the box above the arrow

2 The Sun; chlorophyll; chloroplasts; sugar; glucose; respiration; starch; cellulose; proteins (The last three terms can appear in any order)

3 Reactants are $6CO_2$ and $6H_2O$; products are $6O_2$ and $C_6H_{12}O_6$. The reaction is driven by light energy – in the box above the arrow

4 a Cell membrane: controls what enters and leaves the cell (it allows gases and water to pass in and out of the cell freely, but is a barrier to other chemicals)

  b Cell wall: gives the cell support. It lets water and other chemicals pass through freely

  c The cytoplasm is the jelly-like material that fills the cell, and is where most of the chemical reactions in the cell occur

  d A mitochondrion is a structure in the cell responsible for the release of energy by aerobic respiration/ responsible for the release of most of the energy by the cell

  e Nucleus: contains DNA, which stores the genetic code. The genetic code carries information the cell uses to make enzymes and other proteins

---

5

	Bacteria	Yeast
Outer layer of cell	Cell wall	Cell wall
Genetic material	As circular DNA in the cytoplasm	In the nucleus (as chromosomes)
Respiration	Enzymes for respiration associated with cell membrane	In mitochondria (with some in the cytoplasm)

Page 105 Providing the conditions for photosynthesis

1 The movement of molecules from an area of high concentration to an area of low concentration

2 Carbon dioxide; into; diffusion; oxygen; out of; passive; increase

3 a The chip in the distilled water increased in mass, and the chip in the concentrated sucrose solution decreased in mass. The potato cells contain a dilute solution (in their vacuoles), and the water concentration in the cells is lower than in the distilled water; so the water moves in by osmosis. In the concentrated sucrose solution, the water concentration is lower than in the potato cells; so water is lost by osmosis

  b Cut a number of potato chips and place them in a range of solutions with known concentrations of sucrose. Calculate the percentage change in mass. Repeat the experiment; and calculate the mean change in mass. Plot a graph of percentage change in mass over sucrose concentration. Read off the sucrose concentration where the line crosses the x-axis

4 The concentration of nitrates is higher in the root cells than in the soil, so they cannot be taken up by diffusion. They must be taken up against a concentration gradient; by active transport, which requires energy

5 a As the light intensity increases, the rate of photosynthesis increases; as light energy is required to drive the process. At a certain point, the graph levels off, so any further increase in light intensity will result in no further increase in photosynthesis. At this point, some other factor must be limiting, e.g. carbon dioxide

  b i The graph for the high carbon dioxide concentration has an identical gradient, but reaches a greater height, i.e. photosynthesis reaches faster rate, before levelling off)

   ii In a higher concentration of carbon dioxide, the graph will continue to a higher point (i.e. a higher rate of photosynthesis); until it levels off. At this point (with light and carbon dioxide being available), some other factor (e.g. temperature); must be preventing any further increase in the rate of photosynthesis

Page 106 Fieldwork to investigate plant growth

1 By examining the leaf and answering a sequence of yes or no questions; e.g. does the leaf have needles/is the leaf a typical shape/ are the leaves in groups; Ruby will be able to place the leaf in smaller groups; until she identifies the tree it is from

2 a i 5 cm × 5 cm; ii 0.5 m × 0.5 m; iii 0.5 km × 0.5 km

  b 6 dandelions per m² / 5.6 dandelions per m² (the total number of dandelions over the 10 quadrats is 14, so the mean is 1.4 per quadrat; each quadrat is 0.25 m², so the distribution is 5.6 per m²) (1 mark for answer; 1 mark for units)

  c When the plants show an obvious change in distribution across a location

3 a Light is needed for photosynthesis; and products of photosynthesis are required to synthesise the molecules required for growth; in low light intensities, plants will not be able to photosynthesise; but the tolerance of low light intensities will vary from plant to plant, so some are better able to live in shade than others

  b This evidence supports/increases confidence in the hypothesis; and a mechanism relating to photosynthesis could account for these; but correlation does not prove cause; other factors could also contribute, e.g. competition among the plants for water and minerals

Page 107 How do living things obtain energy?

1 c; e

2 Glucose, oxygen; carbon dioxide, water, energy

3 a 1 mark for reactants; 1 mark for products

   $$C_6H_{12}O_6 + 6O_2 \longrightarrow 6CO_2 + 6H_2O + energy$$

  b The reaction takes place as a series of stages/the equation is a summary; with energy being released in stages

4 Human muscle cells during vigorous exercise; plant roots in waterlogged soil; bacteria in deep puncture wounds

5 a Ethanol/alcohol; carbon dioxide; lactic acid

b Yeast – beer/wine/other alcoholic drink/bread or bacteria – yogurt

6 Aerobic respiration requires the presence of oxygen; anaerobic respiration takes place in the absence of oxygen, or in very low oxygen concentrations. The products of aerobic respiration are carbon dioxide and water; the products of anaerobic respiration vary/ products include alcohol, carbon dioxide, lactic acid, but not water. The energy released by aerobic reaction is much greater than that released by anaerobic respiration

Page 108 B4 Extended response question

5–6 marks

Explains how the sucrose is an exact fit to the enzyme in terms of the active site, protein structure and sequence of amino acids. Explains enzyme specificity and that maltose will not be an exact fit to the active site, and uses an accurate diagram to illustrate these principles. Explains the importance of pH in enzyme action, and the effects of an inappropriate pH on the structure of the active site. Recognises that the optimum pH for invertase must be around 4.5, but the pH optima for other enzymes involved in brewing must also be around this pH. All information in answer is relevant, clear, organised and presented in a structured and coherent format. Specialist terms are used appropriately. There are few, if any, errors in grammar, punctuation and spelling

3–4 marks

Explains enzyme action in terms of the enzyme and lock and key mechanism, but explanation is incomplete, not fully detailed, or related to sucrose. States that enzyme action is specific to one substrate and that enzymes work best at a specific pH, and relates these to enzyme shape, but the answer does not go beyond this. For the most part the information is relevant and presented in a structured and coherent format. Specialist terms are used for the most part appropriately. There are occasional errors in grammar, punctuation and spelling

1–2 marks

Limited description of enzyme action. States that the enzyme works on sucrose only and at a certain pH, with little or no explanation of this. Answer may be simplistic. There may be limited use of specialist terms. Errors of grammar, punctuation and spelling prevent communication of the science

0 marks

Insufficient or irrelevant science. Answer not worthy of credit

B5 Growth and development

Page 109 How organisms develop

1 a Multicellular; specialised

b Tissue; nerve cells; nervous tissue

c Tissues; organs; brain

2 After the eight-cell stage of the embryo, cells become specialised. This is called differentiation. In these specialised cells, only the genes needed to enable the cell to function as that type of cell remain switched on; other genes are switched off

3 Plant tissue: phloem; xylem. Plant organ: flower; root; stem

4 a, b

5 Meristem at tip of shoot/stem: division of meristem cells, followed by enlargement of one of the daughter cells, produces an increase in height/length of stem, or growth of new leaves or flowers. Meristem in side bud: division of meristem cells, followed by enlargement of one of the daughter cells, produces side growth, or growth of new leaves or flowers. Meristem along the length of the stem/shoot and root: division of meristem cells, followed by enlargement of one of the daughter cells, produces an increase in girth/thickness of the stem and root. Meristem in tip of root: division of meristem cells, followed by enlargement of one of the daughter cells, produces an increase in the length of the root (1 mark for each)

Page 110 Plant development

1 Roots grow at the base of the stem; while the shoot continues to grow. The technique enables people to produce many new plants from a single plant

2 Description of taking a cutting or tissue culture:
 • Taking a cutting – cut a small length of plant stem which includes a meristem; dip the cut end into hormone rooting powder; put the end of the stem into damp compost
 • Tissue culture – remove a small piece of tissue, or a few cells from a plant; place on agar; containing nutrients and plant hormones

3 Cell division; cell enlargement

4 Light is coming from one direction, so the plant grows towards the light to expose more surface to the light. This helps the plant's survival by enabling it to photosynthesise. Without photosynthesis, the plant would not be able to grow (as it produces glucose, from which the molecules needed for growth are produced)

5 One direction; towards; phototropism; photosynthesis

6 There is no significant difference between the amount of auxin in the plants kept in the dark or light, or total auxin in plants illuminated on one side; so light has no effect on the *production* of auxin. About 71% of the auxin in the plant illuminated from one side is on the dark side; so as the total auxin was unaffected by light, the auxin must have been *redistributed* from the light to dark side

Page 111 Cell division

1 a A type of cell division; that produces two cells that are genetically identical; and have identical numbers of chromosomes as the parent cell

b During growth; and when cells divide to repair tissues

2 Percentage of total time spent in mitosis: stomach – 2.2%; small intestine – 3.9%; large intestine – 1.3%

3 a The cell increases in size; the number of organelles increases; the DNA in each chromosome is copied

b Rat intestine (30/2000 = 0.15 or 1.5%)

c The developing fruit fly egg; because the egg is developing, so is undergoing rapid cell divisions

4 a, c

Page 112 Chromosomes, genes, DNA and proteins

1 Chromosomes; DNA; genetic; amino acids; cytoplasm

2 a Phosphate (green circle); bases (white rectangles); sugar (yellow pentagon)

b An alpha-helix / like a twisted ladder

3 The genetic code carries the instructions for protein synthesis

4 c, b, e, d

Page 113 Cell specialisation

1 a, d

2 Damaged / diseased; diseased / damaged; adult; limited

3 Michael's first sentence is correct. So for many people, their use is unethical, and is sufficient to prevent their use. Ahmed's first statement is also correct, but many people think that any individual – even an early embryo – has the right to life. Beatrice's statement is correct in that most embryonic stem cells currently come from embryos surplus to IVF treatments, but it is very controversial. Many consider that embryo use is justifiable under any circumstances, but work with, and use of stem cells, is subject to legislation in many countries. Maia is incorrect. While adult stem cells have the potential to replace some cell types, this is much less than that of embryonic stem cells

4

Source of stem cells	One advantage	One disadvantage
Embryo	Can be used to produce any cell type	Removal of stem cells involves destruction of an embryo
Adult	Can be removed from the patient	Used to produce a limited number of cell types only
Therapeutic cloning	Stem cells are genetically identical to those of the patient, so won't be rejected	The 'embryo' produced is still destroyed as stem cells are extracted
Transformed body cells	Potentially, could be used to produce any cell type	None (although the technique is only in its early stages of development)

Page 114 B5 Extended response question

5–6 marks

States that auxins are plant hormones that regulate plant growth, and explains that auxin promotes cell elongation (and cell division) in a plant, so is involved in the plant's growth response to light (phototropism). States that auxin is produced by the tip of the shoot and produces growth below the tip. Describes how, when a plant is exposed to light from one side, auxin is redistributed away from the light to the shaded side, where it produces growth. The shoot therefore grows towards the light. Describes how this is an advantage to the plant because the plant needs light energy for photosynthesis, in order to produce the materials for growth (and energy). All information in answer is relevant, clear, organised and presented in a structured and coherent format. Specialist terms are used appropriately. There are few, if any, errors in grammar, punctuation and spelling

3–4 marks

Describes that the plant grows towards the light because there is more auxin on the shaded side, but the explanation is incomplete and not fully detailed. States that the plant grows towards the light, and describes how, as light is needed for photosynthesis, this is important for the plant to stay alive, but the answer does not go beyond this. For the most part the information is relevant and presented in a structured and coherent format. Specialist terms are used for the most part appropriately. There are occasional errors in grammar, punctuation and spelling

1–2 marks

States that the plant grows towards the light, but there is limited or no description of the action of auxin. States that light is essential for the plant to live, with little or no explanation of this. Answer may be simplistic. There may be limited use of specialist terms. Errors of grammar, punctuation and spelling prevent communication of the science

0 marks

Insufficient or irrelevant science. Answer not worthy of credit

B6 Brain and mind

Page 115 The nervous system

1 a Nervous; hormonal

  b The nervous system uses electrical impulses/messages, the hormonal system uses chemical messages. The nervous system produces a quick, short response, while the hormonal system produces a slower response, but the response is longer-lasting. The nervous system sends messages using nerve cells or neurons, while in the hormonal system, hormones are transported in the blood *(Any 2)*

2 Sensory, receptors, stimuli, central; Motor, central, effectors, muscles

3 Insulin – produced by the pancreas; Oestrogen – produced by the ovaries. It is a sex hormone that controls the development of the adult female body at puberty and the menstrual cycle

4 a Dendrite; Myelin (fatty) sheath; Cell body; Nucleus; Axon

  b Arrow is from left to right

5 Temperature – a higher temperature (but not higher than the body temperature of mammals and birds) speeds up transmission. The diameter of the axon – the wider the axon, the faster the speed. The myelin sheath – the presence of the sheath increases the speed of transmission

Page 116 Linking nerves together

1 b, c

2 a Work in different areas of the body; work between nerves and other nerves, and nerves and muscles; some excite nerves or muscles, different ones inhibit them

  b The receptors on the second nerve or muscle are a specific shape to receive each type of chemical transmitter

3 Stimulus; receptors; organ; ear/eye; eye/ear

4 a Central

  b Hormones, enzymes

  c Contraction, stimuli, heartbeat

Page 117 Reflexes and behaviour

1 a Eye/ receptor; Sensory neuron; Relay neuron in CNS/ brain; Motor neuron

  b It helps to protect the eye from damage if it's suddenly exposed to a bright light

2 Picking up a hot object normally sets up a reflex action where you would drop the plate; when a message reaches the brain that you have picked up the plate, the brain sends a message to motor neurons; which instead of causing you to release the plate, make you hold on to it

3 a 20%; 80%; 70%; 30%

  b Woodlice move towards dark; and wet places

  c Instinctive

  d It assists their survival; since they cannot learn from experience

4 As it has happened many times, the dog has learned to associate the ringing of the doorbell with the arrival of a stranger/someone at the door, so will bark, anticipating the arrival of the stranger

5 A predatory bird, at some stage, will have tried to eat a wasp and will have been stung/harmed in the process. The bird will have come to associate the yellow and black pattern of the wasp with danger; so will avoid insects with similar patterns, such as hoverflies

Page 118 The brain and learning

1 a Cerebral cortex; cerebellum; medulla/brain stem

  b Intelligence; memory; language; consciousness

2 Different types of MRI can investigate the structure and activity; of the brain by scanning it. Other techniques involve stimulating the brain using electrodes and monitoring how the person responds. This requires access to the brain by removing part of the skull; and is usually carried out during brain surgery; so opportunities will be more limited

3 Intelligence; memory; language; consciousness

4 Neurons, links, neuron pathway; repeating, impulses, stimuli

5 It is easier to establish new neuron pathways in children than adults

6 Babies are born with simple instinctive behaviours, e.g. the rooting and sucking reflexes; but then need to learn behaviours. This requires exposure to new, appropriate stimuli. These enable many neuron pathways to develop in the brain. There is evidence to suggest that these pathways, and therefore learning, only develop at a certain age; feral children brought back into society, for instance, develop only limited language skills

Page 119 Memory and drugs

1 a The storage; and retrieval; of information

  b Long-term

  c Arranging them so as to produce a pattern; repetition (of reading, reciting, etc.); other sensible answer *(Any 2)*

2

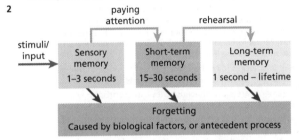

Stimulus; Paying attention; Rehearsal; Sensory, short-term, long term; 1-3 seconds, 15-30 seconds, 1 second-lifetime *(1 mark for stimulus; 1 mark for each arrow; 1 mark for each box)*

3 a Justine; the word (mnemonic) that she makes up is the stimulus

  b Lucas; he is storing information **and** retrieving it

4 Drugs; toxins

5 Bungarotoxin blocks the acetylcholine receptors of the muscles, so the transmitter substance acetylcholine is unable to bind with the receptors on the muscles affected. These muscles are therefore unable to contract, and the person is paralysed

6 When a nerve impulse reaches the end of a neuron, a chemical transmitter is released, which causes the next nerve to fire. After the nerve impulse has passed, the transmitter is taken back up into the neuron and therefore removed from the synapse. MDMA prevents this re-uptake and therefore increases the levels of serotonin in the synapse. The person's mood is lifted. Some time after taking MDMA, the brain becomes depleted of serotonin, and the person feels irritable and tired

Page 120 B6 Extended response question

5–6 marks

Describes in detail the experiments that Pavlov carried out. Uses the terms primary (the food) and secondary stimulus (the bell). Describes that, after a period, the dogs will come to salivate in response to the ringing of a bell, and defines this as a conditioned reflex, where the response has no direct connection with the stimulus. All information in answer is relevant, clear, organised and presented in a structured and coherent format. Specialist terms are used appropriately. There are few, if any, errors in grammar, punctuation and spelling

3–4 marks

Describes in outline the work that Pavlov carried out, and defines this work as being on conditioned reflexes, but misses out some points, e.g. that salivation is the normal response of dogs to the smell, sight or taste of food. States that the dogs came to associate the ringing of the bell with food, but the explanation and terminology used are incomplete. For the most part the information is relevant and presented in a structured and coherent format. Specialist terms are used for the most part appropriately. There are occasional errors in grammar, punctuation and spelling

1–2 marks

There is limited description of the experimental work, and states that the dogs linked the ringing of the bell with food, but with little or no explanation of conditioned reflexes. Answer may be simplistic. There may be limited use of specialist terms. Errors of grammar, punctuation and spelling prevent communication of the science

0 marks

Insufficient or irrelevant science. Answer not worthy of credit

B7 Further biology

Page 121 How the body moves

1 a Support; protection; blood production; movement
   b The muscle has to contract, which pulls the bone
2 a In muscles, when one muscle contracts, the other relaxes; and vice versa
   b When an athlete exercises it increases the size of the muscle so that the muscle can now contract; with a greater force
3 This joint is a hinge joint. The muscles in the arm act in antagonistic pairs; when the biceps contracts, the triceps relaxes and so the forearm is lifted; when the biceps relaxes, the triceps contracts, which brings the arm down again. The muscle only contracts by about 10%; but the joint is a lever and so magnifies the distance moved by the muscle

Page 122 Joints

1 Shoulder; elbow; hips; knee; fingers
   (Any 2)
2 Synovial membrane = produces synovial fluid; Synovial fluid = reduces friction in the joints; Articular cartilage = stops end of bone wearing away; Ligaments = join ends of bones together
3 The role of ligaments is to connect the bones in a joint together; ligaments are elastic, which means they do stretch a little; tendons connect the muscle to the bone; this means that the muscle can transfer its force to the bones; tendons are not elastic, they attach the muscle to the bone
4 Ball and socket joints have a much wider range of movement; hinge joints only allow movement in one plane
5 Ligaments are elastic because they can absorb the energy from an impact; keeping the joint in place
6 Although the risk of injury is high from playing rugby, it does not mean that those risks are not worth taking; there are many benefits to playing sports, including building up muscle and being involved in exercise; although there appear to be a very low number of injuries with watching TV and playing computer games; people who pursue these activities are more likely to become unfit, which may have a larger impact; furthermore, the table does not give an indication of how severe the injuries actually are

Page 123 Exercise and health

1 a Triston = $\frac{115}{(1.9)^2}$ = 31.9 (to 3 s.f.); Spencer = $\frac{100}{(1.85)^2}$ = 29.2 (to 3 s.f.)
   b Triston = obese, Spencer = overweight
   c Spencer is a boxer and this means that he has a lot of muscle, which increases his weight; the only reason it appears he is overweight is his increased muscle mass
2 As Ruth has had surgery she may be on medication that could mask damage when she is exercising; the physiotherapist would also need to know her family history because that could indicate inherited issues
3 This is to prevent excessive strain; on the heart / muscle / joints
4 a A person's weight varies throughout the day; this means that the weights measured will not be as useful as one that is only taken once a day at the same time
   b Taking three weight readings allows Graham to calculate an average; which would be a better estimate of his weight

Page 124 Exercising safely

1 a How often the heart beats per minute
   b The pressure of the circulating blood on the blood vessels
2 glucose + oxygen; → carbon dioxide + water (+ energy released)
3 a 220 − 19; = 201
   b 70%; to 90%
   c Between 1 and 6 minutes
4 a This is where the ligaments overstretch and so the joint becomes more wobbly
   b This is where the ligament of a tendon becomes disconnected
   c This is where the bone in a joint comes out of its socket
5 The treatment for a sprain is PRICE; this means to **protect** the joint from further injury; **rest** the joint so it is not stressed any further; **ice** to cool the injured site down; **compression** to reduce the swelling and prevent further movement; **elevation** to help reduce the swelling
6 Devise a recovery programme suitable for the patient; monitor their recovery programme; asks questions about the history of the patient prior to a programme
   (Any 2)

Page 125 How the heart works

1 a B
   b The blood flows through the heart twice for every circulation of the body
2

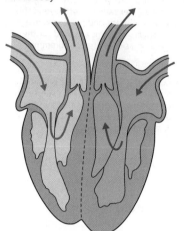

3 a The purpose of the valves is to prevent the backflow of the blood; blood moves to the lung to collect oxygen; faulty valves mean that the blood can remain in the heart
   b The patient may be tired / low on energy / breathless; as the blood that is being pumped around the body is lower in oxygen than it would be normally

Page 126 Blood components

1 Ticks should appear next to: hormones; plasma; red blood cells; platelets; white blood cells
2 Packed with haemoglobin; has no nucleus
   (Any 1)
3 a To keep the blood at the right concentration
   b Carbon dioxide to the lungs, urea to the kidneys
   c Glucose / carbohydrates; proteins / named protein; hormones / named hormone; amino acids; fatty acids; glycerol
   (Any 3)
4 The capillary bed enables efficient transfer of the blood components to and from cells; oxygen and dissolved food moves out of the blood into the tissue fluid; and waste material, such as urea and carbon dioxide, move back into the blood; hormones move to and from the blood; the excess fluid passes into lymph vessels

Page 127 Blood as a transport system

1 a Haemoglobin
   b Left column = vein, right column = artery
2 a To prepare the body for vigorous action / fight or flight
   b To help control blood sugar levels
3 a The cells are bi-concave, which maximises surface area for oxygen absorption; they are round and flexible to enable it to move through the blood vessels easily
   b The sickle cells will now be more likely to clump together; and cause a blockage to blood flow

Page 128 Keeping cool

1 D, A, C, B, E
   (All correct for 3 marks, three correct for 2 marks, two correct for 1 mark)
2 Sodium and potassium are needed for muscle; and nerve function
3 a The astronaut sweats so that when the sweat evaporates it removes some of the excess heat from the body; cooling him down
   b The suit is a closed system; and drinking the lost fluid ensures that he does not become dehydrated
4 a During vasodilation blood flows closer to the surface of the skin; the blood capillaries near the surface dilate; this is so that the hot blood flows closer to the surface; allowing the heat to leave the body at the surface
   b The skin would look red and feel warmer; as the blood is flowing closer to the surface of the skin
   c During vasoconstriction a shunt vessel causes blood to flow further from the skin; the blood capillaries near the surface constrict; this is so that the hot blood flows closer to the core of the body; allowing the heat to keep the core at the correct temperature
   d The skin would look bluish and feel cooler; as the blood is flowing closer to the core than the skin

Page 129 Keeping warm

1 T, T, T, F, F, F

   (All correct for 3 marks, four correct for 2 marks, three correct for 1 mark)

2 a Sensors in the skin; and in the brain

   b Muscles in the skin start to shiver; the cells respire extra sugar

3 a 35°C

   b If warming is too quick, the blood may leave the core; to be replaced with much cooler blood; this could cause heart and organ failure; and possibly death

4 a Hypothalamus

   b An effector is the tissue or gland that carries out a response; they are antagonistic because muscles shiver when the temperature drops too low and this increases the body temperature; once the temperature gets to a high level the shivering stops; sweating acts by starting when the body gets too hot and causes the body to cool down again

Page 130 Diabetes

1 a Insulin

   b Glucose / sugar

   c Kidneys; eyes; circulation

   (Any 2)

2 a Wiremu has to inject insulin because his body no longer produces it; without insulin his body cannot control blood sugar levels

   b Atu controls his diabetes by carefully controlling his diet; and exercising; he controls it this way because, unlike Wiremu, his body still produces insulin; it just does not react to it any more

   c The more sweetened drinks consumed, the higher the risk

3 When blood glucose levels are high, insulin is released into the blood; which causes cells to take up sugar from the blood; when the blood glucose levels drop, insulin production by the pancreas is switched off; this is an example of negative feedback; high glucose in blood causes insulin release, which reduces the glucose level

Page 131 Cycles in nature

1 a It will be incinerated or used as landfill

   b Open loop

   c The resources that made the object can be recycled and so are not lost to landfill; they can be used again in the cycle

2 Plants photosynthesise taking in carbon dioxide, forming sugar; the plants are eaten by animals and the sugar used in respiration in the animal

3 Microorganisms; bacteria; fungi

   (Any 1)

4 Crude oil formed from the bodies of plants and animals; millions of years ago; now humans are taking the crude oil and burning it; the carbon dioxide released will have a big impact on the atmosphere; this is too much for the plants to use to return in a closed loop system; this means the loop is effectively open

5 Microorganisms can obtain energy from the bodies of dead plants and animals; the waste is broken down into simpler forms; that other organisms can then use; for example, the proteins can be broken down into nitrogenous compounds; which help the plants to grow; as well as the animals that feed on them

Page 132 Cycles in rainforests

1 It holds seeds that should grow into a new plant; and it helps ensure dispersal of seeds, as it is attractive to animals such as birds, which eat it

2 a They have a high energy level compared to other parts of a plant

   b They need a lot of energy over a long period, to develop

3 It may cost a lot to transport; a tree may only produce a small number of fruit / supply and demand; there may have been a poor crop so less fruit available

   (Any 2)

4 a The cashew nuts are produced by the tree with the prospect that they will grow into a new tree; the tree produces thousands of nuts because a large number will not be successful and will not grow into a new tree

   b Failure of seed germination; damage by fungi / insects; animal feeding

   (Any 2)

5 a Animal migration; movement of minerals out of the forest; climate change

   (Any 2)

   b Humans are disrupting the cycle in the rainforest by logging; which is removing trees; removing fruits and other resources; that would normally grow into new plants; manufacturing products using rainforest resources; and building homes on rainforest land

   c It is an ethical dilemma, because in the developed nations people have already used a lot of their natural forests; using them as a resource; the people in the rainforests need to survive; it would be difficult for us to argue that they could not use the forest in the ways that we did in our past

Page 133 Protecting soil

1 minerals; air spaces; water; organic material (Four correct for 2 marks, three or two correct for 1 mark)

2 Roots hold the soil together; leaves / stem slows down the rain hitting the soil; remove water from the soil

   (Any 2)

3 a Area = B; the turbidity is high because the trees that would have absorbed the minerals / prevented the soil from being washed away

   b $\frac{10}{0.1} \times 100 = 10\ 000\%$

4 a Supporting; regulation; cultural

   b Removing water for our use; using wood (for building / fuel); using resources for food / medicine

   (Any 2)

Page 134 Ecological services

1 a Carlos

   b Sandy

2 The soil from the rainforest has low levels of minerals because the vegetation grows so fast that the minerals are removed rapidly; soils in the UK, such as on fields, have a higher mineral content because there is less vegetation reliant upon it

3 Once the rainforest has gone, it's gone; the ecological services need to be managed ensuring that the services are still able to support the organisms in the rainforest; this means managing the steady replacement of resources; e.g. replanting trees and not removing too many; this would enable organisms to continue to live

Page 135 Poisoned lakes

1 Bags take a long time to break down in landfill

2 The population involved is low, so the sea can cope

3 a Proteins

   b Anaerobic respiration

4 a Bioaccumulation

   b $\left(\frac{40}{0.04} [1]\right) \times 100; = 100\ 000\%$

Page 136 Sustainable fishing

1 F, T, F, T (Four correct for 2 marks, three or two correct for 1 mark)

2 a As the salmon are grown in nets it is possible that some of their food will escape into the surrounding water; and feed other organisms; this can then alter the natural food chains in the region

   b If the area being farmed was very low then the sea will probably be able to absorb the impact of the farm

3 a This is because, below a certain size, the fish are not yet adults; to ensure that the stocks of fish are replenished we need fish to reach a breeding age; and breed before they are caught

   b This is the term given to the amount of fish in an area old enough to produce eggs

   c This is to prevent people using it as an excuse to get around the quota rules

Page 137 Producing protein and penicillin with microorganisms

1 F, T, T, F, T

   (All correct for 3 marks, four correct for 2 marks, two or three correct for 1 mark)

2 a Diane

   b Wayne; and Ngaire

3 The bacteria must respire using oxygen; so this has to be supplied; as the bacteria respire they will produce carbon dioxide and water as waste products; the carbon dioxide and excess water has to be removed; the bacteria reproduce at an optimum temperature; and pH; this needs to be controlled and adjusted

Page 138 Genetic modification

1 a Genetic modification / engineering

   b The gene for beta carotene is not found in rice plants normally

2 a If you do not produce insulin you would die very quickly; insulin is the only treatment for someone with Type 1 diabetes

   b Some religions treat cows as sacred or pigs as unclean; if you were vegetarian or vegan before 1982 you had to use the product of an animal in your body

3 A, C, E, F, I

   (Four correct for 2 marks, three or two correct for 1 mark)

Page 139 Genetic testing

1 a A complementary sequence of DNA means that the bases in the complementary strand are the opposite to the target strand; this is because A always binds with T; and G always binds with C

   b Complementary strand: TCGCAT

   c If the cells were not washed, there would probably be DNA probe that had not attached to a matching sequence; this would then mean that the unattached probe would still fluoresce under UV light; which could give a false positive

2 The possession of certain genes only indicates an increased probability of having the disorder; there are other factors involved, such as those from the environment; if insurance companies had access to the test results it could mean that you would have to pay higher insurance; in fact you may never actually get the disease; whereas someone who paid less because they did not carry the genes could end up with the disease

Page 140 Nanotechnology

1 a 0.1 to 100 nm

   b $\frac{80\ 000}{2}$; = 40 000

2 a Arwen

   b Claude

   c The nanoparticles block the entry of oxygen, which would cause the lettuce leaves to deteriorate more quickly

3 Set up plates of bacteria; then place a set size of silver bandage on top; in another petri dish put a piece of normal bandage; the same size as the silver one; on top. Leave for a set time; If the bandage was successful at preventing bacterial growth, the bacteria should die; the bacteria on the petri dish which has the normal bandage (the control) should remain

Page 141 Future medicine

1 a Replacing the heart with a transplanted heart

   b There is a shortage of hearts available for donation; and there is less chance of rejection

2 a Signals; electrical; undifferentiated; differentiated

   b There is an ethical issue with using an embryonic stem cell as it has come from an embryo which did not develop into a human; as IPS cells are created from adult body cells, it removes this ethical objection

3 a Leukaemia takes place when the cells that make blood cells becomes cancerous; the treatment involves killing the stem cells in the patient's bone marrow; through radio and chemotherapy; once all the cancerous cells are killed, the stem cells from a bone marrow donor are introduced; these then supply the patient with new blood

   b The bone marrow produces all the blood, including the white blood cells; white blood cells are involved in fighting infections; if they are dead then the infections can occur without any defence

Page 142 Extended response question

5–6 marks

Answer makes clear the arguments for and against genetic testing, including two reasoned points for each argument

Possible marking points about advantages:

- diseases can be anticipated and treatments prepared in advance to prevent onset (forewarned is forearmed)
- should be a decrease in wasteful / ineffective treatment therefore saving money / reducing unnecessary suffering
- people could have earlier tests for the cancer if they carry the higher risk genes, therefore being treated earlier
- screening would be earlier for those at higher risk, therefore any cancer would be picked up earlier with a higher survival rate
- we could know the actual number of people with the risky genes and therefore apportion more accurately NHS funds

Possible marking points about disadvantages:

- some diseases may not be treatable and so knowing earlier could ruin your life
- there may be a false negative and so a person would not be screened for a cancer when they needed it (ignore discussion about false positive as this would lead to increased screening, which is not detrimental to the patient)
- just because you carry certain genes it does not necessarily mean you will get the cancer, therefore the testing may be misleading

All information in the answer is relevant, clear, organised and presented in a structured and coherent format. Specialist terms are used appropriately. There are few, if any, errors in grammar, spelling and punctuation

3–4 marks

The answer makes clear one argument for and one argument against giving a reasoned argument for each point

For the most part the answer is relevant and presented in a structured and coherent format. Specialist terms are used for the most part appropriately. There are occasional errors in grammar, spelling and punctuation

1–2 marks

The answer gives only one side of the argument, or the argument is not clear. The answer may be simplistic. There may be limited use of specialist terms. Communication is impeded by errors in spelling, punctuation and grammar

0 marks

Insufficient or irrelevant science. Answer not worthy of credit